Yousef Nazirizadeh

Photonic crystal slabs for low-cost biosensors

Photonic crystal slabs for low-cost biosensors

by
Yousef Nazirizadeh

KIT Scientific Publishing

Dissertation, Karlsruher Institut für Technologie
Fakultät für Elektrotechnik und Informationstechnik, 2010

Impressum

Karlsruher Institut für Technologie (KIT)
KIT Scientific Publishing
Straße am Forum 2
D-76131 Karlsruhe
www.ksp.kit.edu

KIT – Universität des Landes Baden-Württemberg und nationales
Forschungszentrum in der Helmholtz-Gemeinschaft

KIT Scientific Publishing 2010
Print on Demand

ISBN 978-3-86644-559-8

Photonic crystal slabs for low-cost biosensors

Zur Erlangung des akademischen Grades eines

DOKTOR-INGENIEURS

von der Fakultät für
Elektrotechnik und Informationstechnik
der Karlsruher Institut für Technologie (KIT)
genehmigte

DISSERTATION

von
Yousef Nazirizadeh (M.Sc.)
geb. in Teheran

Tag der mündlichen Prüfung: 21. Juli 2010

Hauptreferent: Prof. Dr. Ulrich Lemmer
Korreferent: Prof. Dr. Martina Gerken
Korreferent: Prof. Dr. Juerg Leuthold

To Mahsa

Luck is what happens when preparation meets opportunity.
SENECA THE YOUNGER

Publications

(1) **Y. Nazirizadeh**, U. Lemmer and M. Gerken, "Colorimetric refractometer based on a two-dimensional photonic crystal slab," *submitted* (2010).

(2) **Y. Nazirizadeh**, U. Bog, S. Sekula, T. Mappes, U. Lemmer and M. Gerken, "Low-cost label-free biosensors using photonic crystals embedded between crossed polarizers," *Opt. Express* **18**, 19120-19128 (2010).

(3) **Y. Nazirizadeh**, U. Lemmer and M. Gerken, "Experimental quality factor determination of guided-mode resonances in photonic crystal slabs," *Appl. Phys. Lett.* **93**, 261110 (2008).

(4) **Y. Nazirizadeh**, U. Geyer, U. Lemmer and M. Gerken, "Spatially resolved optical characterization of photonic crystal slabs using direct evaluation of photonic modes," Proc. *Optical MEMs and Nanophotonics 2008*, 112-113 (2008).

(5) **Y. Nazirizadeh**, J. G. Müller, U. Geyer, U. Lemmer and M. Gerken, "Direct observation of photonic modes in photonic crystal slabs," *Proc. ICTON 2008* **2**, 72-75 (2008).

(6) **Y. Nazirizadeh**, J. Müller, U. Geyer, D. Schelle, E.-B. Kley, A. Tünnermann, U. Lemmer and M. Gerken, "Optical characterization of photonic crystal slabs using orthogonally oriented polarization filters," *Opt. Express* **16**, 7153-7160 (2008).

(7) M. Gerken and **Y. Nazirizadeh**, "Spontaneous emission control for communications devices - a review", *J. Nanophoton.* **2**, 021795 (2008).

Submitted patents

(1) **Y. Nazirizadeh**, U. Lemmer and M. Gerken, "Optisches Element, Verfahren zum Nachweis des Vorhandenseins von Substanzen und Verwendung des optischen Elements," European Patent submission 09005905.6 (29.04.2009).

(2) **Y. Nazirizadeh** and M. Gerken, "Vorrichtung und Verfahren zum Nachweis biologischen Materials," German patent submission DE 10 2010 005 859.9 (26.01.2010).

(3) **Y. Nazirizadeh** and M. Gerken, "Vorrichtung und Verfahren zur Kontrasterhöhung in der Mikroskopie," German patent submission DE 10 2010 005 860.2 (26.01.2010).

CONFERENCE PRESENTATIONS

(1) **Y. Nazirizadeh**, U. Geyer, U. Lemmer and M. Gerken, "Spatally resolved optical characterization of photonic crystal slabs using direct evaluation of photonic modes," *International Conference on Optical MEMS and Nanophotonics*, Freiburg / Germany (2008).

(2) **Y. Nazirizadeh**, J. G. Müller, U. Geyer, U. Lemmer and M. Gerken, "Direct observation of photonic modes in photonic crystal slabs," *International Conference on Transparent Optical Networks*, Athens / Greece (2008).

(3) C. Gärtner, **Y. Nazirizadeh**, C. Karnutsch, S. Uebe and U. Lemmer, "Rate coefficientsfor bimolecular singlet exciton annihilation processes in organic semiconductormaterials," *International Conference on Optical Probes of* π*-Conjugated Polymersand Functional Self Assemblies*, Turku / Finland (2007).

(4) **Y. Nazirizadeh**, J. G. Müller, R. Bornemann, G. Bastian, M. Gerken, U. Lemmer, D. Schelle, E. B. Kley and A. Tünnermann, "Distribution and Emission Properties of Fluorescing Nanospheres on 2D Photonic Crystal Slabs," *CLEO/Europe*, Munich / Germany (2007).

(5) **Y. Nazirizadeh**, J. G. Müller, R. Bornemann, G. Bastian, M. Gerken and U. Lemmer, "Spontaneous Emission Properties in 2D Photonic Crystals," *International Conference on Transparent Optical Networks*, Nottingham / England (2006).

Contents

Chapter 1

Introduction

Summary

Photonic crystals are structures with various facets. Potential applications range from the efficiency enhancement of optoelectronic devices to biosensors. This chapter discusses the background of these structures and gives a motivation for low-cost sensing applications. Furthermore, the structure of the thesis is summarized.

1.1 Motivation

When Eli Yablonovitch and Sajeev John established the term *photonic crystal* with their publications in 1987 [1, 2], it was not the first time that scientists desired to manipulate light with periodic dielectric structures. 99 years earlier, in 1888, Lord Rayleigh investigated periodic stack structures [3], which are known today as one dimensional photonic crystals. These structures show, for instance, a high reflectivity for a specific spectral range and are used today in many applications such as, Bragg-mirrors, reflective coatings or vertical-cavity surface-emitting lasers (VCSELs) [4].

Until today, a variety of photonic crystals have been theoretically proposed and experimentally verified. *Photonic crystal slabs* (PCS), for instance, show light confinement in all three dimensions, although they show a periodicity only in two dimensions.

Applications based on photonic crystals generally use the dispersion relation of modes provided by these structures. Hence, typical applications are slowing of light [5] and the superprism effect [6, 7] utilizing the envelope of modes in the band structure, inhibition of light propagation in the band gap or sensing applications using the evanescent part of photonic crystal modes [8].

PCSs, in particular, are structures, which can be fabricated in an easy and cost efficient way. These structures are composed of a high index layer with the periodic nanostructure, which is surrounded by materials with a lower refractive index. Moreover, these structures offer special modes, called *guided mode resonances* (GMRs) [9, 10], which can be easily addressed via far field measurements and thus significantly simplify its handling. This opens up the way for PCSs to be utilized in sensing applications, for instance, in biotechnology.

Biotechnology is a broad term, which is applied to areas from industrial biotechnology to environmental biotechnology. Medical biotechnology, however, has a special role, as it covers technologies in the pharmaceutical industry and diagnostics. In drug discovery, the biomolecular interactions of molecules are of high interest. Monitoring these interactions in real time delivers important information about the molecules itself.
On the other hand label-free biosensing can greatly improve the diagnostics of diseases [11, 12]. For instance, detection of biomarkers has improved the diagnostics of cancer [13].

Currently, most biological tests require large stationary laboratories and highly qualified technical staff. A common way to determine concentrations of substances in a biological fluid, for instance, is to use an immunoassay such as an enzyme-linked immunosorbent assay (ELISA). Performing ELISA based tests, in general, a long recipe containing washing and injection processes with a final signal analysis has to be followed. A new kind of device, which is the *biosensor*, overcomes these problems and allows for a rapid test under field conditions.

The International Union of Pure and Applied Chemistry (IUPAC) proposes the following definition for biosensors [14, 15]: "A biosensor is a self-contained integrated device, which is capable of providing specific quantitative or semi-quantitative analytical information using a biological recognition element (biochemical receptor), which is in direct spatial contact with a transducer element. A biosensor should be clearly distinguished from a bioanalytical system, which requires additional processing steps, such as reagent addition. Furthermore, a biosensor should be distinguished from a bioprobe, which is either disposable after one measurement, i.e. single use, or unable to continuously monitor the analyte concentration."

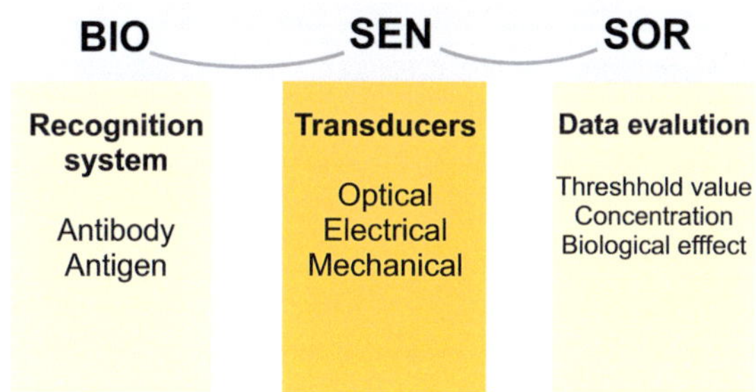

Figure 1.1: *Three components of a biosensor: recognition system, transducer and data evaluation.*

Using this definition a biosensor can be divided into three parts (Fig. 1.1) [15, 16]:

- **The recognition system:** The expression *bio*, in biosensor, originated from its recognition system, which is a biological system. Usually this biological system can bind to a biological substance (molecule, cell, or pathogen) or a chemical analyte and can change its properties, such as refractive index or electrical properties.

- **The transducer:** The task of the transducer is to convert changes of the recognition system into a detectable signal. Depending on the employed physical principle, transducers are classified into three groups - mass sensors, electrochemical sensors, and optical sensors [17]. Mass sensors use a piezoelectric element or a cantilever to detect any presence of biological material. Electrochemical sensors measure the current, resistance or capacitance of the transducer [18, 19]. Optical sensors detect a refractive index change or the presence of a layer [8, 20–28].

- **The data evaluation:** Finally, the detected signal is analyzed regarding any predefined thresholds or concentrations. Moreover, biological effects can be identified from the real-time signal.

Such biosensors show, in particular, promise in simplifying clinical diagnostics, food analysis, bioprocess and environmental monitoring [29, 30].

One of the biggest advantages of biosensors is their approach in label-free sensing. The transducer in such a system adopts the functionality of the label and makes the substance of interest detectable. The transducer, however, works faster, simpler and more physiologically than its labeled counterpart [31]. Moreover, the marker itself, which is usually a dye molecule, quantum dot or radioactive material, can also affect the bioactivity of the labeled substance. The marker, for instance, can provoke misfolding, reduce the mobility or steric hindrance of the substance of interest. In addition, handling with markers is a pricey issue and some of them are incompatible with in-vivo experiments.

Another powerful feature of biosensors is their capability of monitoring biological processes in real-time. The binding kinetic of two molecules, for instance, is of high interest for biochemical and biological research in industry and academia. These biological processes can be monitored using biosensors with transducers based on electrical or mechanical properties. Today's most established label-free approaches, however, are optical methods, in particular, approaches based on surface plasmon resonances (SPRs) [32–38].

SPRs, also know as surface plasmon resonance polaritons, are bounded modes at a dielectric-metal surface, which show an evanescent part within the dielectric. This evanescent wave can probe the surface and changes of its refractive index influence its properties. If the recognition element, which is the functionalized surface, detects any interaction with the substance of interest the spectral position of the SPR is changed in real-time and can be read out. Systems based on this technique require mostly a prism and sophisticated read out optics, which makes them costly and bulky devices.

Alternatively, biosensors can be utilized in flexible point-of-care testing (POCT). These tests are compact and usually user friendly devices for diagnostic tests, which are not performed in a laboratory, but in the hospital at the patient's bed,[1] in the medical practice or in drugstores. A POCT may also be used by the patient himself. Here prominent examples are glucose tests (based on an enzyme reaction) or pregnancy tests (which is in principle an immunoassay).

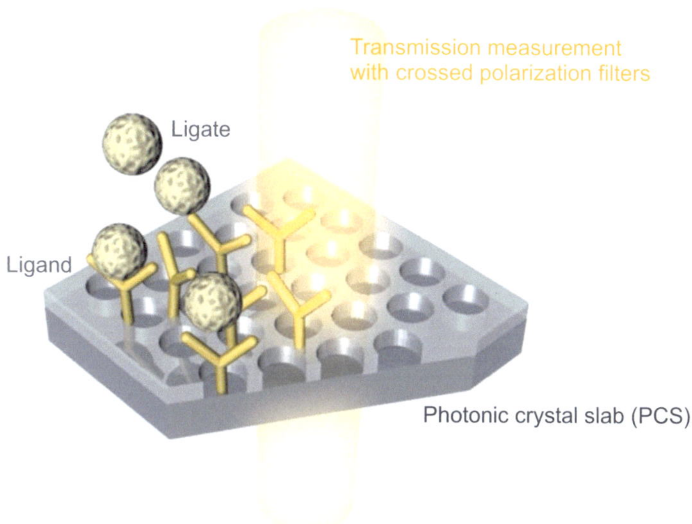

Figure 1.2: *Binding process schematics of a functionalized PCS with a ligate. The ligand is immobilized on the PCS's surface. The binding of the ligand with the ligate induces a refractive index change of the PCS surface, which results in a GMR shift.*

The emphasis of this work is to utilize PCSs as sensors, in particular for biosensing applications. We present a novel and easy-to-implement technology platform for biosensors based on functionalized PCSs. In combination with two crossed polarization

[1] These POCTs are also called bed-side-tests.

filters, a transmission measurement reveals GMRs provided by these PCSs. These modes exhibit an evanescent wave fraction outside the PCS, which probes changes on the surface. As depicted in Fig. 1.2, the binding of the ligate to the immobilized ligand may change the surface refractive index, which leads to a shift in GMR's properties. The appropriate readout procedure allows for compact and cost-efficient biosensors based on this technology platform.

The proper characterization of the fabricated PCS as well as the characterization of provided modes is essential for the design of such biosensors. The homogeneity over the whole structure, for instance, plays an important role to guarantee sensitive sensors.
In the scope of this work we present two biosensor types based on the proposed platform: an optoelectronic biosensor and a visual biosensor. These label-free biosensors are particularly characterized by properties such as compactness, cost-efficiency and user friendliness. Moreover, a method is presented, which allows significant contrast enhancement in light microscopy, in particular for transparent tissue.

1.2 Structure of this thesis

In chapter 2, the basics of photonic crystals and PCSs, in particular, are discussed. The band structures of different photonic crystals are compared and the modes analyzed. GMRs, which are key elements in this work, are discussed in detail in this chapter, too.
Chapter 3 deals with the simulation method, which was used in the scope of this work for the theoretical treatment of transmission measurements, namely the finite-difference time-domain (FDTD) method. Here, a method is discussed, with which only GMRs can be obtained. In this chapter, the impact of geometrical parameters of PCSs is discussed as well.
In chapter 4, an optical configuration is introduced, which allows the direct view to GMRs, while suppressing the unwanted background light. For this purpose, two orthogonally oriented polarization filters are used.
In chapter 5, PCSs are characterized regarding their homogeneity, which is an essential issue, when utilizing PCS for sensor applications. Furthermore, spectral position and Q-factor of GMRs provided by such structures are the matter of investigation.
In chapter 6, an optoelectronic biosensor is introduced, which is capable of detecting the binding process of two molecules in real time. In this chapter, this capability is demonstrated using the binding of streptavidin to biotin. The characteristics of this sensor is its simplicity, which results in a compact and cost-efficient device.

A visual sensor, which can be realized as a compact and cost-efficient device, too, is the topic of chapter 7. Here the ambient light serves as the light source, while the human eye is the detector.
In chapter 8, the PCS is utilized for contrast enhancement in light microscopy, specially for transparent tissues. Here two different ways of how the contrast can be enhanced are discussed.
In chapter 9, the further perspectives of such sensors are discussed.

Chapter 2

Photonic crystal slabs

Summary

In the scope of this thesis a series of novel sensor types based on photonic crystal slabs is presented. One of their most important characteristic is their potential for a cost-efficient realization. In this chapter, photonic crystals and photonic crystal slabs, in particular, are in focus. Moreover, the physical background as well as the most important applications and fabrication methods are discussed.

2.1 Photonic crystals - an overview

Photonic crystals are mostly artificial materials consisting of one or more dielectric materials. These dielectrics are spatially arranged in a way that the refractive index is repeated periodically and has a periodicity in the range of the photon's wavelength to be manipulated. Photonic crystals as known today have been described for the first time by E. Yablonovitch [1] and S. John [2] almost simultaneously in 1987. At that time, they intended to develop a material that has similar effects on photons as semiconductors on electrons. In Fig. 2.1, the comparison between an electron in an atomic potential and a photon in a periodic structure is depicted. Analogous to the mechanisms in a semiconductor, in photonic crystals, interfaces of refractive index changes act as centers for the scattering of photons. Such scattering centers are also often referred to as "dielectric atoms". The scattered waves starting at these centers can interfere constructively or destructively and result in standing waves in the structure. The stationary properties of these standing waves, in terms of their longitudinal or transverse energy distribution, are called *modes*. In multi-dimensional photonic crystals, the energy-momentum relation of these modes often results in complex relations. To illustrate the dispersion relation one uses, as in crystallography and solid state physics, the so-called *band structure*, which plots the energy of the photon against its wave vector. Analogous to semiconductors, under certain conditions, such as a sufficient refractive index contrast between the alternating dielectrics, a band gap can arise in the band structure. In such band gaps, also called *stopbands*, the formation of modes is inhibited. In other words, a photon with a wavelength in the band gap cannot propagate in the photonic crystal.

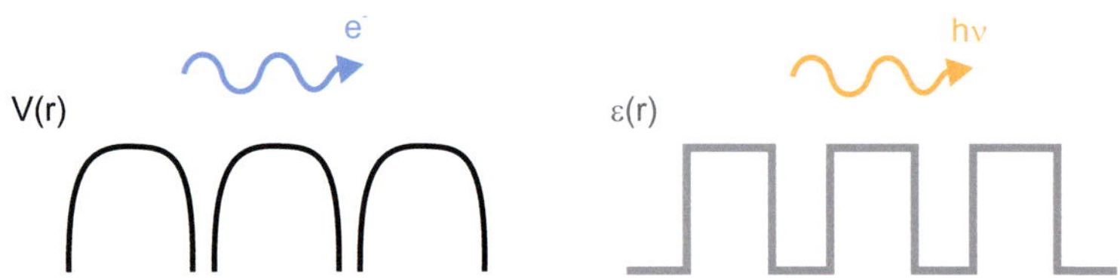

Figure 2.1: *Analogy between electron waves in a periodic potential, and light waves in a periodically varying dielectric environment.*

Unlike in solid state physics, the transitions between the electron levels or bands are not of major importance, but rather the existence or non existence of discrete modes and their dispersion relation. Photonic crystals therefore allow not only traditional applications, such as designing dichroic mirrors, interference filters, or dispersion compensators, but also quantum mechanical manipulation or the targeting of directional characteristics by

emitters introduced in such crystals.
Due to the linearity of Maxwell's equations, which is used to describe the behavior of photonic crystals, the properties of these crystals are scalable to different wavelengths with their size. This is also the reason why experimental achievements were initially performed for microwaves and later for infrared radiation and visible light.

Let us now return to the pioneers of these structures and deal with the question of why they have proposed photonic crystals. Yablonovitch postulated that the spontaneous emission of an emitter can be suppressed by means of a periodic dielectric structure. He argued with the expected band gap, in which the spontaneous emission of electron transitions in semiconductors after Purcell [39] was not allowed. He saw the biggest potential of his work in laser physics, where spontaneous emission has an important influence on the laser threshold, or in solar cells where the internal spontaneous emission could be reabsorbed by the structure. The experimental evidence of the manipulation of spontaneous emission was offered by groups around S. Noda and W. L. Vos [40,41]. In a further investigation it was impressively demonstrated, how spontaneous emission can be manipulated in a much simpler photonic crystal [42].
In contrast to this, John wanted to produce a localization of light with a defect in the periodic structure. This phenomenon was already known from semiconductor physics (Anderson localization). Here again, the experimental proof was delivered relatively late in 1997 [43].

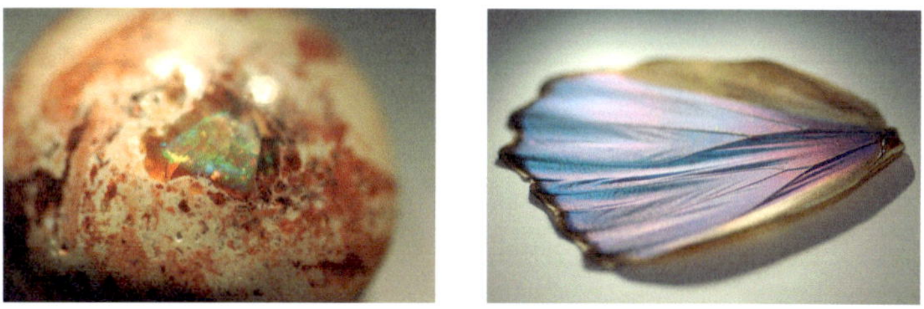

Figure 2.2: *Naturally occurring photonic crystals. Left image: Opal surrounded by stone. Right image: A butterfly's wings.*[2]

Regardless of these considerations there are naturally occurring structures called photonic crystals as well. They are composed, as their artificially fabricated counter parts, of periodic dielectric structures. In the left image of Fig. 2.2 an opal surrounded by stone is shown. Opals consist of silica spheres of about 150 to 400 nm in diameter, which are arranged in a close packing order. This order shows periodicities in the range of visible light in all spatial directions and as a result photonic crystal effects occur, which let the opal appear colorful. In the right image of Fig. 2.2, a picture of a butterfly's wing is shown. Unlike in the opal, this construction exhibits a periodicity in only two dimensions. Nevertheless, it shows reflections of a specific spectral range, which are due to the photonic crystal properties.

2.1.1 One-, two- and three-dimensional photonic crystals

The following provides an overview of the wide variety of different photonic crystals. In general, we distinguish between one-dimensional (1D), two-dimensional (2D), and three-dimensional (3D) photonic crystals. A schematic representation of such structures is depicted in Fig. 2.3.

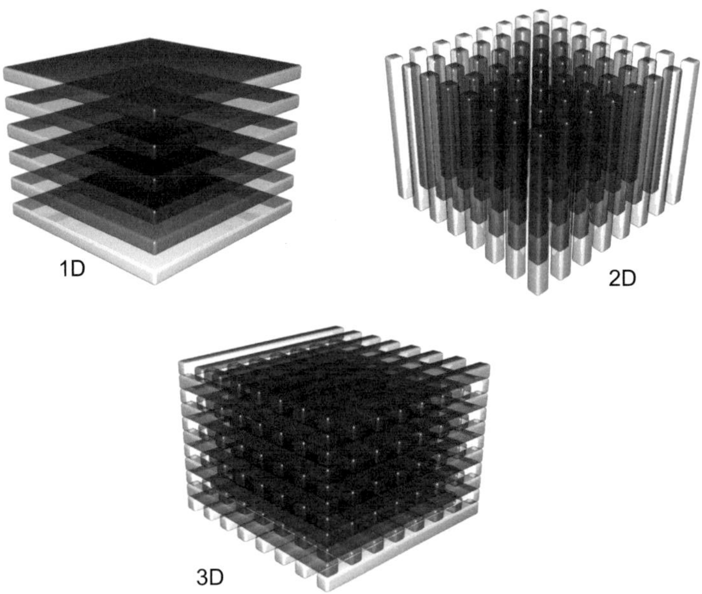

Figure 2.3: *Schematic representation of one-, two-and three-dimensional photonic crystals.*

One-dimensional photonic crystals

Although a photonic crystal can exhibit its periodicity in all spatial directions and photonic crystal effects are a result of complex interference pattern, the basic principles of such structures can be explained in a simple manner on the basis of *one-dimensional (1D) photonic crystals*. The following intuitive explanation also justifies the fact that 1D structures are known and applied for more than 120 years, much earlier than multi-dimensional photonic crystals were considered.

1D photonic crystals are characterized by their periodicity in only one spatial direction. A plane wave propagating in these structures perpendicular to the periodicity undergoes partial reflections at each refractive index change, according to Fresnel equations. For this reason, two waves propagating in opposite direction arise. Now, if the periodicity of the structure is half the size of the wavelength of the plane waves this results in destructive interference for the counter propagating wave. In other words, the plane wave with the wavelength half the periodicity cannot penetrate the structure and is completely reflected. This is known as the 1D Bragg condition:

$$m\lambda_0 = 2n_{eff}\Lambda \qquad m = 1,2,... \tag{2.1}$$

Here, λ_0 is the wavelength of the plane wave, Λ is the periodicity of the structure, n_{eff} is the effective refractive index and m is the order of the Bragg condition. This spectral region of inhibited propagation is also called the *band gap* of the structure.

Similarly, constructive interference occurs at a frequency of $\lambda_0/4$ and the plane wave can traverse the structure without reflection. This plane wave, which is resonant in the structure, is referred as to a *photonic crystal mode*.

Two-dimensional photonic crystals

Two-dimensional (2D) photonic crystals exhibit a variation of the refractive index in two spatial directions. Perpendicular to this plane the dielectric medium is homogeneous. One can think of it as infinitely long rods of one dielectric in another dielectric with different refractive indices. Photonic crystal effects can only be observed, therefore, in the plane with the refractive index variation. Again, here constructive and destructive interferences result in the formation of photonic crystals modes and band gaps. Unlike in 1D photonic crystals, in 2D photonic crystals two polarization directions of the electromagnetic waves are distinguishable. On the one hand, transverse electric modes[3] (TE modes) and on the other hand transverse magnetic modes[4] (TM modes) can propagate in the structure. This distinction applies only to waves with a propagation direction in the plane of periodicity, waves with a different propagation direction show a mixture of the two polarizations and

[3] H-field in the plane of periodicity

[4] E-field in the plane of periodicity

therefore cannot be clearly distinguished.

The most prominent example of 2D photonic crystals is the photonic crystal fiber [44]. This structure is extended almost infinitely perpendicular to the plane of periodicity. The electromagnetic wave is guided in a nucleus, which is surrounded by the photonic crystal. Such fibers are firstly used for low-loss and low-dispersion signal transmission by guiding modes in an air core instead of a glass core. Secondly, these fibers are used to enale high nonlinear processes via guiding modes in the dielectric and locally limited to a small volume. The typical applications of this are supercontinuum generation [45] and fiber lasers.

Three-dimensional photonic crystals

Three-dimensional (3D) photonic crystals show a variation of the refractive index in all three spatial directions and allow for manipulation of light in all directions in space. Therefore a complete band gap for all polarizations and all spatial directions can be enforced. For this reason, the first experimental evidence of spontaneous emission suppression was based on these kind of photonic crystals [41].
3D photonic crystals are realized in many variations. The first structure to offer a full band gap was the Yablonovite [46]. This structure consists of three pillars of holes that are etched in a dielectric at angles of 35.26° and 120°, relative to each other. As these crystals, due to their elaborate fabrication, could not be further miniaturized to obtain photonic crystals in the visible spectrum of light, new fabrication methods had to be developed. Today the most popular examples are *woodpile photonic crystals* and opal structures [47]. Opals occur, similar as in nature, under appropriate sedimentation of colloids, which constitute a fcc[5] lattice and thus have a periodic structure in all spatial directions. The size of the colloids, mostly consisting of SiO_2, determine the periodicity of the structure. These photonic crystals are distinguished by their simple manufacturing of crystals for the visible spectral range. Woodpile photonic crystals, on the other hand, require more sophisticated fabrication methods. They are mostly fabricated using a direct laser writing approach [48].

2.2 Photonic crystal slabs

One of the great advantages of photonic crystals is the possibility to manipulate electromagnetic radiation in all spatial directions. The pioneering work of E. Yablonovitch [1]

[5]face centered cubic

and S. John [2] addressed 3D structures, which also offer light manipulation in all spatial directions.
As discussed in the previous section, 3D photonic crystals are technologically challenging structures. A photonic crystal showing a less complex structure is desirable, not only for cost reasons, but also due to simpler integration to already existing production processes. Integrated optical circuits, for instance, are constructed, similar to the microelectronic circuits, on flat substrates in various manufacturing steps layer by layer (*bottom-up*). The integration of 3D photonic crystals, which should be additionally built up out of the plane, represents a major challenge.

A very good compromise in terms of structure simplicity and light manipulation in all spatial directions are *photonic crystal slabs* (PCSs) [49, 50]. Such hybrid crystals consist of a dielectric layer with a higher refractive index relative to its environment, which exhibits a periodic structure [51]. A PCS can either be a free standing slab in air called *air-bridge* PCS or be realized on a substrate. The latter is mechanically more stable and easier to fabricate.

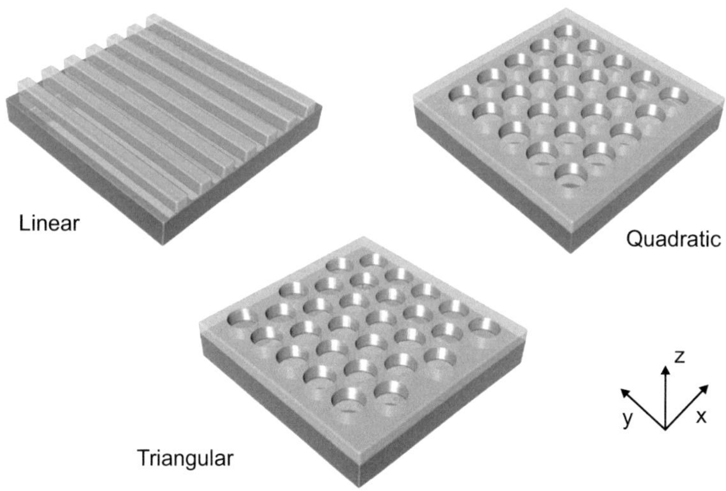

Figure 2.4: *Schematic representation of linear, quadratic and triangular photonic crystal slabs (PCSs).*

Depending on the periodicity of PCSs, these structures can be classified into groups. In Fig. 2.4 the three most important groups, which have been treated in this study, are

schematically shown: *linear*, *quadratic* and *triangular* PCS. Linear PCSs have a periodicity in only one spatial direction (in our example x direction), while square and triangular PCSs show a periodicity in both, x and y direction.
The definition of the space coordinates x, y and z as defined in Fig. 2.4 holds for all PCSs used in this work, except otherwise indicated.

In a simplified picture, one can identify two effects that cause light manipulation in all spatial directions in PCSs. First, the periodic structure, which is introduced in the high refractive index layer. Due to production reasons, the periodic structures are usually either holes in the layer or columns consisting of high index material. This periodicity affects, similarly as in 2D photonic crystals, light manipulation in the plane of the layer. Second, the light, perpendicular to the plane of the layer, is restricted locally partly by total internal reflection and partly by simple reflection. For this reason, PCSs are also referred to as (quasi) 3D photonic crystals.
To understand the detailed properties of PCSs, the band structure of them is presented and provided modes are discussed in the next section.

2.3 Band structure of photonic crystal slabs

In analogy to crystallography or semiconductor physics, we use the dispersion relation, the relation between the energy and the wave vector, to represent the characteristics of photonic crystals. This relation is plotted in graphs called *band structure*. In such graphs, one can relatively easily investigate the dispersion relation of modes and identify band gaps.
In this chapter, methods of calculating and plotting photonic crystal modes are described. Furthermore, the characteristics of PCSs are explained on the basis of exemplary band diagrams.

2.3.1 Maxwell equations

The macroscopic Maxwell's equations represent the physical basis of electromagnetic wave propagation in photonic crystals. These equations are shown here in their differential form in SI units [52]:

$$\nabla \cdot \vec{D} = \rho \tag{2.2}$$

$$\nabla \times \vec{E} = -\frac{\partial \vec{B}}{\partial t} \tag{2.3}$$

$$\nabla \cdot \vec{B} = 0 \tag{2.4}$$

$$\nabla \times \vec{H} = \frac{\partial \vec{D}}{\partial t} + \vec{j}. \tag{2.5}$$

Here, $\vec{E}$ and $\vec{H}$ are electric and magnetic field vectors, $\vec{D}$ is the dielectric displacement, $\vec{B}$ is the magnetic flux, $\vec{j}$ is the current density and ρ represents free charge density.
Before performing further transformations, we want to simplify these equations with the help of certain boundary conditions. First, free charges or currents can be neglected, i.e. $\vec{j} = 0$ and $\rho = 0$. Furthermore, we consider only dielectric media with a relative permeability of $\mu_r = 1$. This assumption is a good approximation[6] for paramagnetic and diamagnetic media, as deviations from unity are generally in the range of $\pm 10^{-5}$. In addition, employed materials are considered to be linear and isotropic.

After some algebraic transformations, and the assumption of harmonic waves for $\vec{E}$ and $\vec{H}$, the physics of Maxwell's equations can be transfered into an eigenvalue problem. The derivation of this equation, known in the literature as the *master equation*, can be found in [52].

$$\nabla \times \left(\frac{1}{\epsilon(\vec{r})} \nabla \times \vec{H}(r) \right) = \left(\frac{\omega}{c} \right)^2 \vec{H}(r) \tag{2.6}$$

Here $\epsilon(\vec{r})$ is the spatially changing dielectric constant. This equation has some similarities to the Schrödinger equation known from semiconductor physics, which is the reason for analogies between electrons in semiconductors and photons in photonic crystals.
Using this equation, field distributions and dispersion relations, i. e., the frequency ω as a function of the wave vector $\vec{k}$, are found for the $\epsilon(\vec{r})$ of interest.
The solution of this equation that is known as the Hermitian eigenvalue problem, delivers harmonic magnetic modes. In a further step the electric fields ($\vec{E}$) will be determined using these magnetic fields ($\vec{H}$).

An interesting property of photonic crystals is their length scalability. In general, there is no fundamental constant for photonic crystals that depends on the spatial dimension of the crystal. This is also the reason for the simple relationship between crystals of various sizes. The solution of the eigenvalue problem for a unit of length also provides

[6] For ferromagnetic media $\mu = 1$ does not hold anymore.

the solutions of other units of length. This behavior can be best explained with the master equation (2.6). When the spatial dimensions of a photonic crystals are scaled by a factor s, i. e., $\epsilon'(\vec{r}) = \epsilon(\vec{r}/s)$, the master equation can be rewritten with $\vec{H}(\vec{r}) = \vec{H}'(\vec{r}/s)$ for the fields and $\omega' = \omega/s$ for the frequency.

2.3.2 The reciprocal lattice

Most photonic crystals show discrete translational symmetries.[7] This means that a spatial shift of the crystal by a certain vector will not change the geometry, which can be written as $\epsilon(\vec{r} + \vec{a}) = \epsilon(\vec{r})$. This translational symmetry allows for the definition of a *primitive cell* for the crystal, which is the smallest volume element defined by a multiple of the primitive translation vectors[8] $\vec{a_1}$, $\vec{a_2}$ and $\vec{a_3}$.

To calculate the band structure of a photonic crystal, the corresponding reciprocal space is used. This is the Fourier transform of the real space, whose primitive vectors are derived from primitive translation vectors as follows:

$$\vec{b_1} = 2\pi \frac{\vec{a_2} \times \vec{a_3}}{\vec{a_1} \cdot (\vec{a_2} \times \vec{a_3})}, \qquad \vec{b_2} = 2\pi \frac{\vec{a_3} \times \vec{a_1}}{\vec{a_2} \cdot (\vec{a_3} \times \vec{a_1})}, \qquad \vec{b_3} = 2\pi \frac{\vec{a_1} \times \vec{a_2}}{\vec{a_3} \cdot (\vec{a_1} \times \vec{a_2})}. \tag{2.7}$$

Similar to the real space, in the reciprocal space an elementary cell can be defined, whereon the solutions of the master equation (2.6) can be restricted. This unit cell is referred to as the first Brillouin zone. Certain photonic crystals, however, exhibit not only translational, but also rotational or mirror symmetries. For this reason, the first Brillouin zone can be further restricted to the irreducible Brillouin zone.

In Fig. 2.5 and 2.6 the two most common geometries of photonic crystals used in this work are shown schematically in two dimensions, namely the quadratic and triangular geometry. The left parts of both figures depict the real space of the crystal, in which, regardless of the unit cell shape, red grid points are plotted in the center of each unit element. These grid points are regain in the right part of each figure, which is the reciprocal space. All wave vectors, which lie outside the Brillouin zone, can be addressed via adding a multiple of the reciprocal lattice vector. A closer look at the grid points bares that after transformation in the reciprocal space, they have a rotation of 90°. This effect can be clearly observed in triangular crystals.

[7]Excluded are photonic quasicrystals, which have only a discrete rotational symmetry.

[8]These are the shortest vectors that point from a lattice point to another.

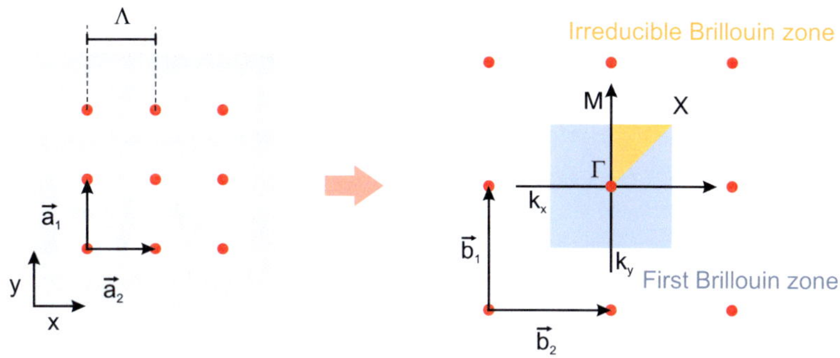

Figure 2.5: *A quadratic photonic crystal in real space and reciprocal space. In the reciprocal space the first Brillouin zone is marked with a blue square. The irreducible Brillouin zone, which constructs the first Brillouin zone completely by rotation and mirroring, is marked with a yellow triangle.*

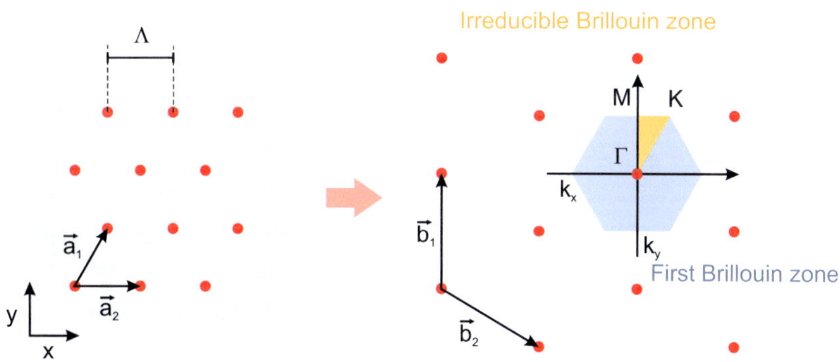

Figure 2.6: *A triangular photonic crystal in real space and reciprocal space.*

2.3.3 The band structure

For band structure calculations, usually the eigenvalue problem (2.6) is solved in frequency space. For each $\vec{k}$ the eigenmodes and eigenvalues are calculated and plotted in a diagram. This diagram, referred to as the *band structure*, displays the wave vector $\vec{k}$ on the abscissa and the normalized frequency a/λ on the ordinate.

Before discussing the band structure of PCSs, we start with the band structure of a 2D photonic crystal. A common method to calculate the band structure of 1D, 2D or 3D photonic crystals is the *plane wave expansion (PWE) method* [53, 54]. In this work, to obtain the band structure of our first example, the PWE method is also applied. The band structure of a 2D photonic crystal, consisting of infinitely long air holes in a homogeneous medium[9] with a triangular geometry (Fig. 2.7 left image), is shown in Fig. 2.7. The ratio of hole radius to the periodicity (r/Λ) is 0.45. To simplify the results, absorption in the dielectric and the frequency dependency of the refractive index (dispersion) are not included in the calculations.

As these structures show their periodicity only in x and y directions, we plot on the abscissa of the band structure $k_{||} = \sqrt{k_x^2 + k_y^2}$ instead of the full vector $\vec{k}$. Thus, these plots are the *projected band structures*. The center of the first Brillouin zone is located on a grid point (Fig. 2.5 and 2.6) and is usually referred to as the Γ point. At this point, the wave vector in the plane of the periodicity, which is $k_{||}$, is zero. The two remaining symmetry points in a square crystal are referred to as M and X point and in a triangular crystal as M and K point.

In 2D photonic crystals one can generally distinguish between two noninteracting modes using their symmetry: the transverse electric (TE) and transverse magnetic (TM) polarization.

- TE: $\vec{E}$ parallel to the air cylinders, $\vec{H}$ in the plane of periodicity
- TM: $\vec{H}$ parallel to the air cylinders, $\vec{E}$ in the plane of periodicity

This polarization decoupling is motivated by the fact that the field in the plane of periodicity results again in purely longitudinal field, after scattering in the structure. And a field, aligned parallel to the air cylinders, still retains its transversality.

For both polarizations independently different dispersion relations and stop bands arise. In these stop bands, the propagation of electromagnetic radiation is prohibited. An overlap of stop bands of the two polarizations is referred to as a *complete band gap*.

The size of the stop bands or the existence of a complete band gap depends in this exemplary 2D photonic crystal on the refractive index contrast of the materials involved and the hole radius. For r/Λ = 0.3, for instance, there is no stop band for TM polarization and consequently no complete band gap.

Due to the fact that the PWE method can only be applied for infinitely extended crystals the guided-mode expansion (GME) method should be applied for obtaining the complete band structure of PCSs [55]. In this method, the electromagnetic fields are represented as

[9] Here the optical properties of silicon at telecom wavelengths is used (n=3.518 @ λ = 1550 nm).

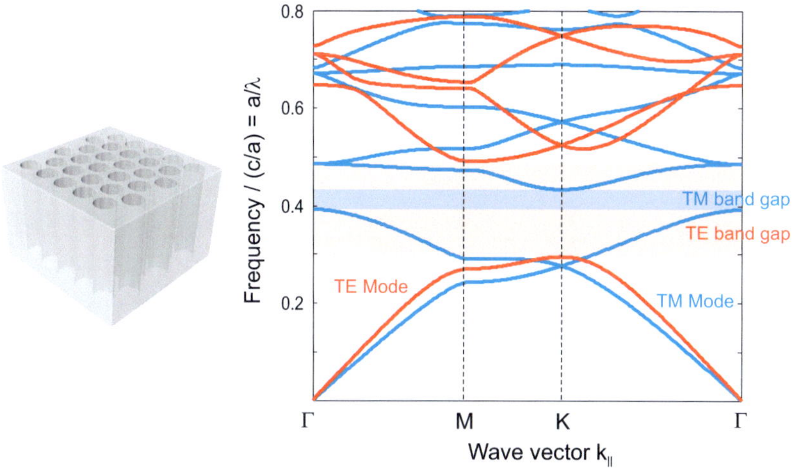

Figure 2.7: *Band structure of a 2D triangular photonic crystal consisting of infinitely long holes in silicon. Stop bands are present in both the TM, as well as in the TE polarization. The overlap of these bands is called a complete band gap.*

a combination of 2D plane waves in the plane of periodicity and guided waves in a perpendicular plane to it. This method brings two important advantages; first, additional boundary conditions, such as absorbing layers, may be omitted in the simulation. Second, the computational cost is relatively low, compared to other alternatives like finite-difference time-domain (FDTD) method.

One disadvantage of this method is that some guided modes are not taken into account, for instance, lossy modes with a component to the far field. Consequently, solutions from this method are only approximations. During this work, however, it was shown that calculated band structures agree well with measured results.

In Fig. 2.8 the band structure of a PCS is calculated applying the GME method. Here, the crystal exhibits the same geometric parameters and material properties as in the previous example, but its extension in the z-direction is only $d/\Lambda = 0.6$. Because of the lack of translational symmetry in the vertical direction a distinct decoupling of TE and TM polarization is no longer possible. Therefore symmetric and antisymmetric modes, referring to the symmetry of the field distribution in z plane, can be differentiated. These modes, however, are dominated by TE and TM polarization [56] and are referred to as even and odd modes, respectively. The lowest-order (fundamental) even mode and the lowest-order odd mode have their electric and magnetic field, respectively, parallel to the slab.

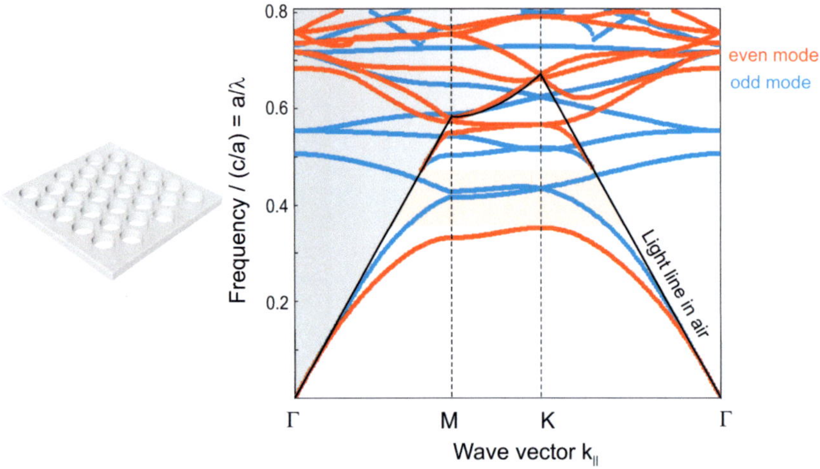

Figure 2.8: *Band structure of a photonic crystal slab, consisting of a silicon layer surrounded by air with holes in a triangular geometry. The gray marked area is the light cone, while the black line represents the light line. Photonic crystal slabs do not have a complete band gap. A band gap exists only for symmetric modes.*

- Symmetric (even) modes: dominated by TE polarization (TE-like)
- Antisymmetric (odd) modes: dominated by TM polarization (TM-like)

In this example, a stop band for the symmetric modes is present. It is generally known that no stop band arises for odd modes in PCS,[10] which means the absence of a complete band gap in such structures.

Modes in a slab waveguide (without any periodic structure) can be classified into two groups. Looking at the equation, which is the total reflection condition in a slab waveguide with the definition of θ from figure 2.9

$$\theta = \arctan\left(\frac{k_{||}}{k_z}\right) = \arcsin\left(\frac{c_0 k_{||}}{n_{core}\omega}\right) \tag{2.8}$$

with $k_{||} = \sqrt{k_x^2 + k_y^2}$ as the wave vector in the plane of the slab waveguide and k_z the wave vector perpendicular to it, a critical angle for total reflection can be derived.

[10]For PCSs composed of holes in a high index layer. For PCSs composed of high index rods, there exists no stop band for the even modes.

$$\theta_{critical} = \arcsin\left(\frac{n_{cladding}}{n_{core}}\right) \tag{2.9}$$

For modes with $k_{||} > n_{cladding}\omega/c_0$,[11] i. e. $\theta > \theta_{critical}$, the total reflection condition is true. Thus, these modes cannot escape from the structure and are guided in the slab. Therefore, these modes are denoted as *guided modes*. Outside of the slab waveguide, the wave vector of these modes becomes complex and shows an exponentially decaying intensity profile. This part of the mode is referred to as the *evanescent wave*.
On the other hand, modes with $k_{||} < n_{mantel}\omega/c_0$ do not fulfill the total reflection condition. Due to reflections based on the Fresnel formula, so-called *quasi-guided modes* occur in these structures. These modes have an intrinsic loss mechanism, which means a part of the mode is coupled to the far field.
The terminology of these modes is not yet established in the photonic crystal community. Quasi-guided modes (QGM) and leaky modes are the most frequently used common terms. In this work, we use the term quasi-guided mode.

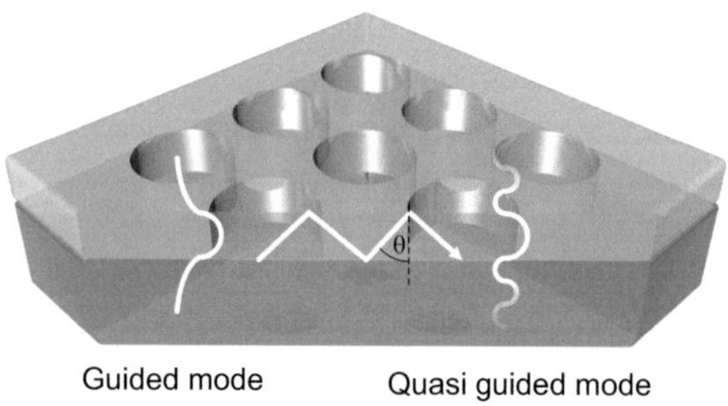

Figure 2.9: *Schematic representation of guided and quasi-guided modes in a PCS.*

The classification of guided modes and QGM is used in PCSs as well. A schematic representation of these modes is given in Fig. 2.9.
As previously discussed, the dispersion relation of a plane wave in the cladding material defines the limit condition for QGM and guided modes. In Fig. 2.8 the dispersion relation in air is drawn into the band structure. Modes below this line, which is referred to as the *light line*, cannot propagate freely in the cladding material and are guided modes, whereas, modes above the light line are QGM. In addition, there exists a continuum of modes above the light line, which is outlined in Fig. 2.8 as a gray area.

[11]This is the dispersion relation of plane waves in the cladding material.

Due to practical reasons, PCSs are often fabricated on thick substrates. Therefore, as a final example, we discuss a PCS on a quartz substrate. In Fig. 2.10 the band structure, obtained using the GME method, of such a PCS is shown. Two key characteristics are different here compared to a structure without substrate.

First, we cannot distinguish clearly between symmetric and antisymmetric modes due to the broken symmetry. Here, one can speak only of quasi-symmetric and quasi-antisymmetric modes.

Second, an additional light line of plane waves in the substrate is drawn in the band structure. Guided modes are then only modes lying below both light lines. Modes between these two light lines are captured in the substrate and cannot escape to the far field.

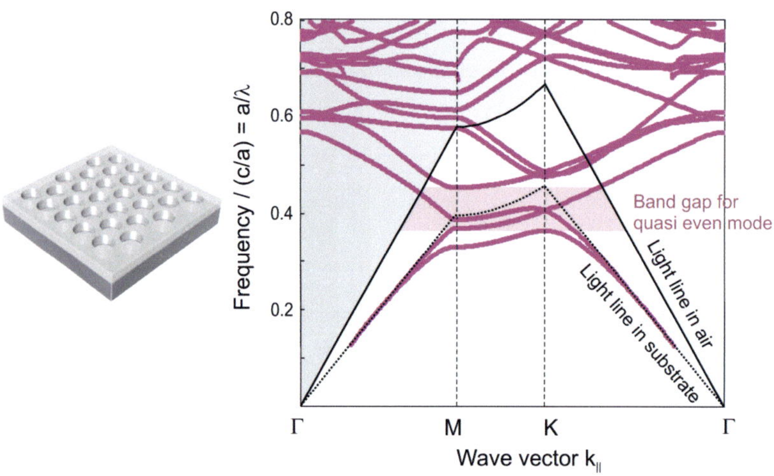

Figure 2.10: *Band structure of a PCS on a quartz substrate, consisting of holes in a Nb_2O_5 layer. Because of the broken symmetry, there are no symmetric and antisymmetric modes, but only quasi-symmetric and antisymmetric modes. A second light line is added in this band structure, which is the dispersion relation of a plane wave in the substrate material.*

2.4 Guided-mode resonances

In the previous section, we discussed the QGMs, which are modes lying above the light line. These modes have an intrinsic loss to external radiation, i. e., while these modes

propagate in the structure, they couple to the far field. Likewise, electromagnetic radiation, incident from the far field, can couple to quasi guided modes.
While performing transmission or reflection measurements on PCSs, a spectral part of the light source will couple to the PCS, depending on the geometrical parameter and the angle of incidence. However, the excited quasi guided mode will radiate to both sides. In the transmission measurement part of the light is missing, whereas in the reflection measurement more light is observed. We refer to these resonances as *guided-mode resonances* (GMRs) [57].

Independent of these considerations, R. W. Wood observed similar resonances in 1902 [58, 59]. He observed rapid variations in the intensity of light sources diffracted by gratings. As he could not model them using classic theories, he named them anomalies in the reflection. No theory could describe *Wood's anomalies* convincingly, until A. Hessel and A. A. Oliner [60] presented a theory based on guided waves in a slab.

Fig. 2.11 shows transmissivity and reflectivity measurements of structured (black line) and unstructured (gray line) surface areas under normal incidence. In the unstructured region, Fabry-Perot oscillations due to the thin-film are clearly visible for both transmission and reflection. In the structured region, however, these thin-film effects are superimposed with photonic crystal GMRs.

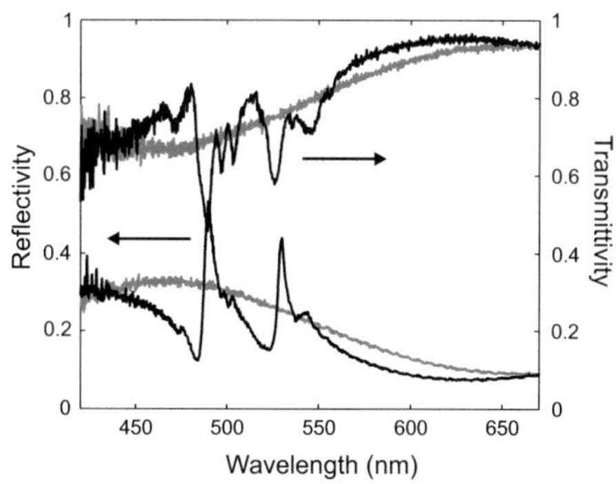

Figure 2.11: *Transmission and reflection measurement on photonic crystal slab. Black lines indicate measurements on structured and gray lines on unstructured regions.*

In solid-state and atomic physics, similar resonances are usually described with the Fano notation [10, 61]. In the case of GMRs, the line shape of the transmitted field amplitude can be described with the following term [62]:

$$t = f_0 \frac{\left(q + \frac{2(\lambda - \lambda_0)}{\gamma}\right)^2}{1 + \left(\frac{2(\lambda - \lambda_0)}{\gamma}\right)^2} \tag{2.10}$$

where λ_0 is the resonance wavelength, γ is the FWHM[12] and f_0 is the strength of the resonance. The factor q varies between +1 and -1 and determines the shape of the resonance. In real transmission measurements, this term is usually superimposed with an *Airy* function, which models the background thin-film interference pattern [62]. Equation 2.10 is used for fitting procedures to find the linewidth of measured resonances.

2.5 Application fields for photonic crystal slabs

In the following, we will discuss application fields of photonic crystals, in particular for photonic crystal slabs. A variety of applications based on photonic crystals have been proposed and experimentally realized, such as band pass filters [9], waveguides [63, 64], slowing of light [5], superprisma effect [6,7], negative refraction and cavities in the crystal structures. However, here we present only the most important applications that are highly relevant to this work, namely: spontaneous emission control, light extraction in *light emitting diodes* (LED) and sensors based on photonic crystal slabs.

2.5.1 Spontaneous emission control

The performance of many optoelectronic devices is related to spontaneous emission. In an LED, for example, a non optimal design will direct the spontaneous emission into waveguide modes, which are trapped in the LED, and reduce their external efficiency [65]. The threshold of semiconductor lasers depends on the fraction of produced light by spontaneous emission into the lasing mode. An ideal laser is a thresholdless laser, in which all spontaneously generated photons are emitted into the laser mode [66, 67]. The unwanted spontaneous emission of excited states in solar cells reduces their efficiency. This is the reason for efforts to control spontaneous emission, either by suppressing or enhancing it.

The pioneering work in this field was done by Purcell in 1946. He suggested placing an emitter in a cavity will cause the spontaneous emission to change in terms of its relaxation time [39]. Whereas, Kleppner proposed in 1981 to use cavities smaller than the emission

[12] Full width at half maximum

wavelength to inhibit the spontaneous emission [68].
Photonic crystal cavities or homogeneous photonic crystals are ideal candidates for this purpose [69], as they provide photon confinement smaller than the emission wavelength. The most simple situation is a dipole point source in the photonic crystal. Treating the spontaneous emission classically,[13] Fermi's Golden Rule states that the spontaneous emission decay rate $\Gamma(\vec{r}, \vec{d}, \omega)$ of an initial state is obtained by summing over all available final states. In an ideal and infinitely extended photonic crystal the *local radiative density of states* (LRDOS) is proportional to the decay rate and is defined as followed

$$LRDOS(\vec{r}, \vec{d}, \omega) = \frac{1}{\epsilon(\vec{r})} \sum_{k} 2\omega\delta(\omega^2 - \omega_k^2) \left| \vec{d_0} \cdot \Psi_k(\vec{r}) \right|^2 \tag{2.11}$$

here, $\vec{r}$ is the position of the dipole emitter, $\vec{d} = d\vec{d_0}$ is the dipole moment and ω its frequency. $\Psi_k(\vec{r})$ are the electromagnetic eigenmodes with corresponding eigenfrequencies ω_k. Furthermore the decay rate can be expressed as

$$\Gamma(\vec{r}, \vec{d}, \omega) = \frac{1}{\tau_{rad}} \tag{2.12}$$

where τ_{rad} is the radiative lifetime.
Considering equation 2.11, the spontaneous emission properties may be controlled by manipulating the number of available modes as well as the position of the dipole relative to the electromagnetic field.

In [71–73] simulations show the position-dependent radiation rate. However, until now this effect could not yet be proven in experiment.
The manipulation of the overall spontaneous emission rate was shown experimentally by Ogawa et al. in [40] and by Lodhal et al. in [41]. A more clear evidence was given by Fujia et al. in [42] using a PCS structure. The lifetime (τ_{rad}), which is anti proportional to the decay rate, was increased by a factor of 5, only by matching the photonic crystal band gap with the emission wavelength of the emitter.

2.5.2 Light extraction in LEDs

An LED usually consists of semiconductor thin films with a high refractive index grown on a substrate. Furthermore in an *organic light emitting diode* (OLED) usually one of the contacts is a metal layer on top of the structure. Therefore the light has to be coupled out through the transparent substrate. One of the main loss mechanisms in LEDs is the poor extraction efficiency.

[13] A discussion of quantum analysis versus classical analysis can be found in [70].

In a standard OLED, for instance, only 20% of the generated light is coupled to the desiered farfield. About 30% are lost due to total internal reflection on the air-substrate interface, while the remaining 50% are trapped in guided modes and *surface plasmon polaritons* (SPP) in the OLED. Many techniques are proposed to couple as much as possible of these bounded modes to the farfield, such as introducing roughness to the involved layers or using microoptics on the substrate. One of the most promising methods is to pattern the layers in the OLED to obtain a PCS, providing quasi guided modes [74–77]. These modes are leaky to the farfield as discussed in section 2.3.3 and may enhance the extraction efficiency of the device. However, patterning these layers may also change the electrical properties of the OLED in a negative way.

2.5.3 Sensors based on photonic crystals

Photonic crystals are a suitable platform for many sensor applications. There is a variety of proposals for detecting biological and chemical substances or simply changes of refractive index of a fluid. All these sensors are based on photonic crystal modes. The evanescent part of these modes is designed to interact with the substance to be detected, which is in direct contact to the photonic crystal. An effective refractive index change on the surface of the photonic crystal will cause a shift of the spectral position of these modes. There are two possible reasons for this effective refractive index change: either a concentration change of a substance in the fluid, or deposition of a substance on the surface. To obtain the latter case the surface of the photonic crystal is functionalized in a way that it can bind biological or chemical materials. For the readout of the changes usually transmission or reflection measurements in a proper direction are used.

Some concepts use cavity modes in photonic crystal cavities with a high quality factor [78, 79]. Two main advantages motivate this concept. First, the high quality factor results in a very sharp resonance, which in combination with a high resolution spectrometer improves the detection sensitivity. Furthermore, the high quality factor allows for a long interaction time of the mode with its surrounding, which increases the detection sensitivity as well. The main drawbacks of this concept are the complex fabrication process and readout technology.

Another concept, which does not have the previously mentioned method's drawbacks, is using the GMRs in a photonic crystal slab to detect refractive index changes. For this purpose, a large area photonic crystal slab can be used, which may be produced in a cost-efficient way [80, 81]. Additionally, a transmission measurement at normal incidence allows to access the modes in a simple way. Cunningham et al. [8, 82, 83] published a series of articles based on this concept. They used the light of an LED to illuminate the photonic crystal slab, while a spectrometer analyzed the reflected light that exhibits the GMRs. Any reaction on the surface of the photonic crystal was followed by a resonance shift. In this configuration the binding kinetics of biological substances with the functionalized surface are shown.

In general a photonic crystal can be used to detect any environmental change on the photonic crystal, which causes a shift in the provided modes. For instance elongation of the crystal will change the periodicity and will cause a shift of the modes [84, 85].

2.6 Fabrication method of photonic crystals

Progresses in nanotechnology recently made the fabrication of photonic crystals for the visible range possible. However, first experiments were done with photonic crystals designed for the infrared region [86].
There are several ways to obtain a periodic dielectric structure. Self-organization, for instance, is used to produce three dimensional photonic crystal structures [87]. A dilution of colloidal spheres of a certain dielectric will form either by natural sedimentation or applied forces an ffc crystal. In the scope of this work, however, we will discuss only the fabrication processes of photonic crystal slabs.

In principle, a photonic crystal slab consists either of a high index layer with holes or of high index material pillars, of course in a periodic order as discussed in section 2.2.
One of the standard technologies is to use lithography to bring the desired pattern into the substrate with an ensuing dry etching process. First, the substrate, with the high index layer, is coated with a thin photoresist layer. A lithography step is than used to expose the photoresists. Due to the importance of this step, more details are given in the next subsections. After development, the samples are coated with a blocking layer, such as chromium. A lift-off process leads to a structured blocking layer, which is transferred into the high index layer by applying physical plasma etching. The structure depth is regulated by the etching time. Finally, the blocking layer is wet-chemically removed.

2.6.1 Electron beam lithography

Electron beam lithography is a technique of exposing a resist with an electron beam. This beam is guided line by line across the sample, thus this technique is applied without a lithography mask. Unlike photolithography, with today's electron optics possible pattern resolution is down to a few tens of nm. This technique, however, has low throughput and high costs, therefore electron beam lithography is mostly used for prototyping or mask fabrication for conventional photolithography.

The scanning field of the electron beam is limited to some hundred micrometers. To expose bigger areas, a mechanical translation stage should be used that can lead to dislocation of the pattern and might cause unwanted stitching errors. Particularly in the case of photonic crystal slabs, these errors will cause disturbances in the periodicity and hence defects in the crystal. This turns out to be a crucial issue.
Moreover, exposing large area photonic crystals, the exposure dose should be stabilized. Otherwise, possible drifts in the exposure dose will influence the geometrical parameters of the crystal. For instance the hole diameter is directly related to the beam current.

2.6.2 Laser interference lithography

Laser interference lithography offers an alternative to electron beam lithography, without the mentioned disadvantages. This method is used to expose large area gratings with a fixed periodicity.[14]
The basic concept is the same as in interferometry. A photoresist is exposed with a superposition of at least two expanded beams of a coherent light source. The interference patterns are periodic series of fringes, consisting of intensity minima and maxima. By selecting the appropriate angle θ of these beams, the interference grating periodicity can be adjusted using $(\lambda/2)/sin(\theta/2)$, where λ is the wavelength of the light source.
To achieve higher order symmetries, the samples are rotated and exposed a second time. For a quadratic grating, for instance, the sample is rotated by 90°, while for a triangular symmetry a rotation of 60° is required.

Beside the disadvantage of being limited to periodic structures, one can face many technological challenges using this method. Unwanted reflections of the light source, for example, can cause macroscopic over gratings, which may lead to a quality reduction.

2.6.3 Nanoimprint lithography and injection molding

An alternative ways to produce nano structures are *nanoimprint lithography* and injection molding. Unlike the previous methods, here no light exposure for the lithography step and no etching process is required. Hence, these methods reduce fabrication steps and have the potential for a low-cost and high throughput photonic crystal fabrication method.

[14] There exist methods to fabricate chirped gratings that is gratings with changing periodicity along the substrate.

Nanoimprint lithography was first proposed by Chou et al. [88] in 1996. There are two important techniques: thermoplastic and photo nanoimprint lithography. A conventionally produced mold, with the desired nano structure, is pressed either in a thermoplastic polymer under a certain pressure and heat, or into a photo curable liquid resist with an ensuing UV light exposure.

Injection molding is a standard manufacturing process, which is deployed in a variety of products. In general, a thermoplastic or thermosetting plastic material is injected into a mold cavity, where it cools and hardens to the desired shape. This technique allows for producing parts with macroscopic and microscopic features in one step, such as a petri dish containing a nano structured surface.

To obtain a PCS, a high index layer on top of these nano structures is required. This needs to be coated in a further step.

2.7 Suitable materials for photonic crystal slabs

In this work sensor types are presented, which are based on transmission properties of photonic crystal slabs, mostly in the visible spectral range. The materials used for these crystals have to fulfill some requirements, such as mechanical stability and good transmission properties.

The refractive index is the most important property of materials for photonic crystals. The refractive index contrast at the interface of the periodic structure causes the scattering of light resulting in the formation of modes. In literature the symbol n is commonly used and it is defined as $n = \sqrt{\epsilon_r \mu_r}$, where ϵ_r is the material's relative permittivity, and μ_r is its relative permeability. However, for most materials μ_r is very close to 1 at optical frequencies, such as for many paramagnetic and diamagnetic materials. In real materials, two effects are responsible for another important property, the absorption. First the polarization does not respond instantaneously to an applied field, which causes dielectric losses. Second the non-perfect insulator properties lead to absorption losses. Considering these aspects, we can define a complex refractive index $\widetilde{n} = n + i\kappa$, where n is the refractive index and κ the extinction coefficient, indicating the amount of absorption loss. The relation between the real and complex part of the refractive index can be describe using the Kramers-Kronig relations. This complex refractive index is a function of the electromagnetic wave frequency. There are several models to describe this dependency, such as Sellmeier's equation for dielectrics or Drude's model for metals.

Considering PCS at least two materials are involved, namely the slab material and the surrounding material. The refractive index of the slab material should offer a higher value compared to its surrounding. Throughout this work the photonic crystal slab was mechanically supported by a substrate, whose refractive index was chosen to be smaller than the slab material. The extinction coefficient κ of all participating materials should be as small as possible, firstly allowing transmission measurements and secondly preventing mode absorption. The absorption of provided modes will shorten the absorption length, which has the following consequences for a sensor based on photonic crystal slabs:

- lower quality factor for the mode, i.e. a broader linewidth, resulting in a lower resolution in some sensor applications
- shorter interaction time of the mode with the surrounding material, resulting in a lower sensitivity

In table 2.1 an overview of materials used in this work is presented. Note that varying fabrication parameters and processes may lead to high variations in the complex refractive index of these materials.

	n	κ	Part
ITO	2.0	0.01	slab
Ta_2O_5	2.07	0.0009	slab
Nb_2O_5	2.32	0.0023	slab
Glas	1.46	≈ 0	substrate
Quartz	1.55	≈ 0	substrate
Air	1.000268	≈ 0	surrounding

Table 2.1: *Overview of refractive index and extinction coefficient of used materials. All values are given at 500nm.*

Indium tin oxide (ITO), tantalum pentoxide (Ta_2O_5) and niobium pentoxide (Nb_2O_5) are adopted as the high index slab materials. These materials are well known in other technologies. ITO, for example, is a transparent and robust anode for the display technology, Ta_2O_5 is used widely in optical coatings and Nb_2O_5 is a well known superconductor. Sputtering or thermal evaporation can be used for coating the desired slab on a substrate. As a substrate material glass,[15] quartz or poly(methyl methacrylate) (PMMA) can be used. While glass and quartz are mechanically and chemically robust materials, PMMA is a flexible and not always chemically robust material. However, its flexibility makes

[15] Microscope slide used in this work

this material a perfect candidate for thermoplastic nanoimprinting.

In the scope of this work three different PCS classes were used. We refer to them as PCS_{Nb2O5},[16] PCS_{Ta2O5} and PCS_{ITO}.[17] In table 2.2 the used materials, PCS parameters and fabrication methods are summarized.

	substrate material	slab material	slab thickness	type of periodicity	fabrication method
PCS_{Nb2O5}	quartz	Nb_2O_5	150 nm 221 nm	triangular	electron beam lithography
PCS_{Ta2O5}	glass	Ta_2O_5	300 nm	linear	interference lithography
PCS_{ITO}	glass	ITO	135 nm	linear quadratic	interference lithography

Table 2.2: *List of PCSs employed in the scope of this work.*

[16]These PCSs were fabricated at the Institute of Applied Physics at the Friedrich-Schiller-Universität Jena by Detlef Schelle, Ernst-Bernhard Kley and Andreas Tünnermann.

[17]These PCSs were fabricated at the LTI by Ulf Geyer.

Chapter 3

Simulations using the finite-difference time-domain method

Summary

This chapter will look at simulations of transmission through a photonic crystal slab. As the simulation method, the finite-difference time-domain method was chosen, since this method solves the time-dependent Maxwell's equations in a spatial grid and thus all physical features are included. Beginning with the basic concept of this method, we will proceed with the approach of simulating the transmission through a photonic crystal slab. Furthermore, a special filter technique, which allows for a direct access to guided-mode resonances, will be introduced.

At the end of this chapter the influence of the photonic crystal slab's geometrical parameters on the provided modes is evaluated using the mentioned method.

3.1 Finite-difference time-domain method

The emphasis of this work is on the application of photonic crystal slabs to design low cost sensors. For this reason, proper characterization and theoretical treatment of photonic crystals are crucially important. Therefore, during this thesis, selected experiments are accompanied with simulations to verify their physical evidence. The simulation method used for this purpose was the *finite-difference time-domain* (FDTD) method.

The FDTD method was first proposed by Yee in 1966 [89] as a numerical method to solve the discretized time-dependent Maxwell's equations. The term finite-difference time-domain and its corresponding acronym FDTD, however, were established by Taflove in 1980 [90]. In principle this method is based on a *leapfrog* method in a Cartesian grid, with $\vec{E}$ and $\vec{H}$ fields shifted by half a spatial and half a time step. Using a Fourier transform, the solutions in time domain can be transfered into the frequency domain. This property allows with a single run the simulation of a wide frequency range.

Due to insufficient computational power, initially little interest was shown for the FDTD method. However, with increasing computational power, the impact of this method was continuously growing in the last decades. Furthermore, this method is an intuitive and popular tool, which has been implemented to a variety of commercial and noncommercial software solutions.

3.1.1 Physical background

We start with the Maxwell's curl equations 2.3 and 2.5 and consider a linear and isotropic medium, i.e. the electric displacement field is $\vec{D} = \epsilon(\omega)\vec{E}$ and the magnetic field is $\vec{B} = \frac{1}{\mu}\vec{H}$. We end up with six scalar equations in Cartesian coordinates:

$$\frac{\partial H_x}{\partial t} = \frac{1}{\mu}\left(\frac{\partial E_y}{\partial z} - \frac{\partial E_z}{\partial y}\right) \tag{3.1}$$

$$\frac{\partial H_y}{\partial t} = \frac{1}{\mu}\left(\frac{\partial E_z}{\partial x} - \frac{\partial E_x}{\partial z}\right) \tag{3.2}$$

$$\frac{\partial H_z}{\partial t} = \frac{1}{\mu}\left(\frac{\partial E_x}{\partial y} - \frac{\partial E_y}{\partial x}\right) \tag{3.3}$$

$$\frac{\partial E_x}{\partial t} = \frac{1}{\epsilon}\left(\frac{\partial H_z}{\partial y} - \frac{\partial H_y}{\partial z} - \sigma E_x\right) \tag{3.4}$$

$$\frac{\partial E_y}{\partial t} = \frac{1}{\epsilon}\left(\frac{\partial H_x}{\partial z} - \frac{\partial H_z}{\partial x} - \sigma E_y\right) \tag{3.5}$$

$$\frac{\partial E_z}{\partial t} = \frac{1}{\epsilon}\left(\frac{\partial H_y}{\partial x} - \frac{\partial H_x}{\partial y} - \sigma E_z\right) \tag{3.6}$$

From these equations, it can be concluded that the derivative of the $\vec{E}$ field is related to the curl of the $\vec{H}$ field and vice versa. This relation represents the base of the *Yee algorithm.* In Yee's algorithm $\vec{E}$ and $\vec{H}$ are defined as they are shown in figure 3.1. Here the grid coordinates are defined as $(i, j, k) = (i\Delta x, j\Delta y, k\Delta z)$, where Δx, Δy and Δz are the actual grid units.

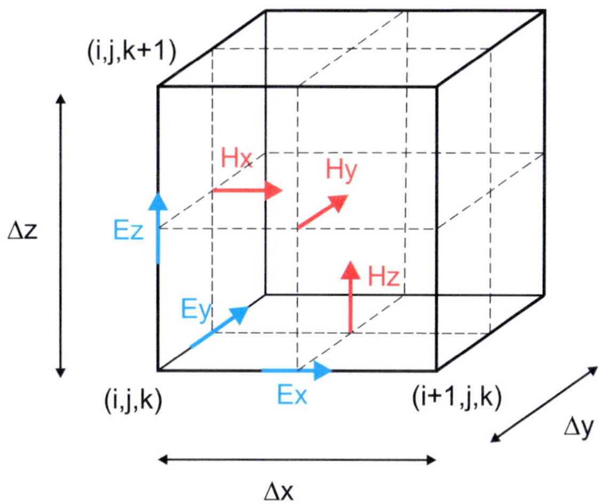

Figure 3.1: *Arrangement of the electrical and magnetic field components in the cubical grid of the Yee cell in blue and red respectively, at the point (i,j,k).*

Furthermore, in the FDTD algorithm the electric and magnetic field components are calculated, shifted by half a time step (Δt) iteratively. The electromagnetic field component at the time step n+1 can be calculated from the electric field component at the time step n and the surrounding magnetic components at the time step n+1/2. Analog to this, the magnetic field component at the time step n+1/2 can be calculated from the magnetic field component at the time step n-1/2 and the surrounding electrical field components at the time step n. This alternately calculation of the electric and magnetic field components is called *leapfrog* method.

For accurate results, certain stability conditions need to be fulfilled. The most important one is the relation between the spatial and temporal step size. The time step is limited by the dimensions of the spatial grid and the velocity of the electromagnetic waves [91]

$$\Delta t \leq \frac{1}{\sqrt{\frac{1}{\Delta x^2} + \frac{1}{\Delta y^2} + \frac{1}{\Delta z^2}}} \frac{1}{c_{max}}, \tag{3.7}$$

where c_{max} is the maximum velocity of waves.

A variety of sources can be used during a simulation to obtain its scattered fields from the structure of interest, such as a Gaussian pulsed plane wave or sinusoidal plane wave. Since in this work the simulation of transmission through photonic crystals for a finite broad spectral range is intended, a Gaussian pulsed plane wave source is used. This source contains a spectrum of electromagnetic waves and thus a single simulation run is sufficient, to obtain the transmission spectra.

In FDTD the electric and magnetic fields are calculated at discrete spatial positions. These values are then used for the calculations of the next time step. The outcome of the simulation, which are these values as a function of the time are recorded after each simulation step. Furthermore the scattered and/or radiated far field can be obtained from these values as reported in [92].

3.1.2 Boundary conditions

To avoid an infinite simulation procedure, the simulation domain needs to be restricted spatially to the region of interest. At the edges of this region the simulation domain is terminated with boundary conditions. A simple example of a boundary condition is the *Dirichlet boundary condition*, which simply sets the solution of the wave equations equal to zero at the edge of the grid. As a consequence all incident waves are reflected back, which corresponds to an ideal metallic boundary. In most of the cases, however, this kind of boundary condition cannot be used. Simulating, for example, an infinite expanded space, a boundary condition with no reflection of the incident waves is needed. In this case, the boundary conditions should have an absorbing character. In simulating periodic structures, such as photonic crystals, a boundary condition should be applied, which is able to repeat the unit cell of the structure .

Perfectly matched layer boundary condition

One of the first attempts to create boundary condition without reflection of the incident wave, was the *absorbing boundary condition* (ABC). It turns out that ABCs work well in one dimension. In two or three dimensions, which may be an important scenario, however, their functionality is limited.

Another approach was the *perfectly matched layer* (PML) boundary condition, which is more an absorption region rather than a boundary condition. This concept was initially formulated by Berrenger in 1994 [93] and provides an angle and frequency independent absorption. A PML boundary condition is composed of multiple artificial absorbing layers placed on the edge of the simulation domain. An entering wave is absorbed by these layers with a minimum of reflection on each layer interface. However, reflections on the starting edge of the PML and at the edge of the terminating wall remain an issue. Another critical point is the angle dependent absorption of the PML [94]. This effect can cause intense problems, particularly when other boundary conditions, such as periodic boundary conditions, are used. In such cases reflections on the PML can cause artificial resonances in the simulation domain.

Bloch-periodic boundary condition

The *periodic boundary condition* (PBC) is useful when both, the structure and the electromagnetic fields are periodic. In a 1D simulation domain with a size of L, for instance, the field components follow

$$f(x+L) = f(x). \tag{3.8}$$

This condition, however, does not include any phase shift between each period. Therefore, a plane wave at non-normal angle of incidence cannot be treated. This issue requires a more generalized condition:

$$f(x+L) = e^{ik_x L} f(x). \tag{3.9}$$

Here, k_x is the wave vector. This condition is known as the *Bloch-periodic boundary condition* (BPBC). As a drawback compared to the PBC, BPBC implicates a higher memory and computation time consumption.

3.1.3 Strengths and weaknesses of finite-difference time-domain modeling

Probably the most important advantage of the FDTD method is its calculation in the time-domain. Using a broadband source, such as a Gaussian pulse, this allows to obtain the system's response over a wide frequency range from a single simulation run. Furthermore FDTD calculates $\vec{E}$ and $\vec{H}$ fields in a physically easy to understand way (solving

Maxwell's equation in time and space), which can be used to generate movies and pictures.

On the other hand FDTD consumes high computational and memory resources. The entire computational domain, for instance, needs to be discretized in a spatial grid. This problem, however, can be limited by choosing the proper boundary conditions. The development process of these boundary conditions is not completed yet and using them still requires operating experiences.

3.2 Simulating the transmission through a photonic crystal slab

In this work we present a technology platform for sensors, in particular for biosensor, based on GMR in PCS. The environmental surrounding of the PCS influences the spectral position of these GMRs. In a transmission measurement the spectral position of the GMR is determined and analyzed. To understand the interaction of the QGMs, which is the origin of the GMRs, with its surrounding and to optimize this interaction, we used FDTD as a modeling tool.

Similar to the measurement, we performed a transmission simulation and obtained a spectrum containing information about the spectral position of the GMR and its line width. For this purpose we used a plane wave Gaussian pulse, which contains the frequency of interest, at various angles. This electromagnetic wave excites QGMs in the PCS. Using field monitors, the transmission, which contains information about these modes, is recorded.

In the scope of this work *FDTD Solutions* a commercially available software offered by Lumerical Solutions, Inc. [95], is applied. FDTD Solutions exhibits a graphical user interface with a computer aided design (CAD) interface. Furthermore a script language, which is comparable to the Matlab script, can be used to realize series of simulations.

Alternatively, *Meep* [96], which is a free software published under the GNU General Public License, was used. This software, however, does not offer a graphical user interface.

3.2.1 Simulation domain

In order to choose the appropriate simulation domain, problem specific requirements need to be properly defined. These requirements define the boundary conditions, the size of the cell, and the required sources and monitors.

In the case of simulating the transmission through a PCS, a plane wave is used as the illumination source (Fig. 3.2). This plane wave generates a short electromagnetic Gaussian pulse that allows for a broadband excitation in a single run. We placed the source in the upper part of the simulation domain with a distance of approximately one grid to the PML. For simulation in the k-space, the wave vector k of this source is varied. We record the transmission using a point monitor, which was placed in the lower part of the simulation domain.

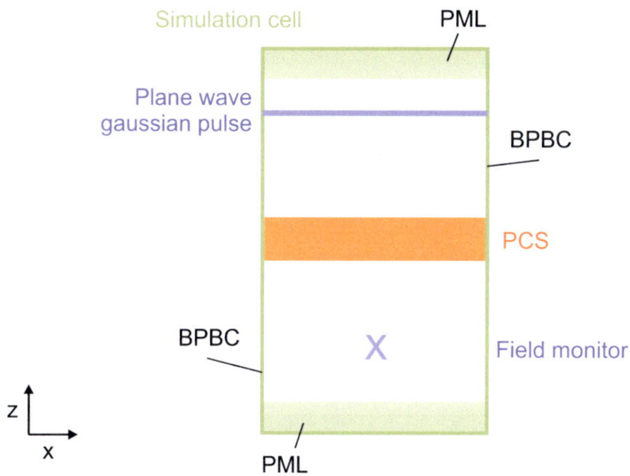

Figure 3.2: *A 2D sketch of the simulation domain used for transmission simulation through a PCS. At the top and the bottom PML absorbing boundary conditions are imposed. For the remaining surfaces we impose BPBC. A plane wave Gaussian pulse is used as the illumination and a field monitor is used to record the transmission.*

For a PCS, which is a dielectric structure with a periodicity in two dimensions, PBC for two dimensions is a straightforward solution. Using a PBC, the structure, the sources, and the monitors are periodically repeated. Hence, to realize only one dipole source, the PBC cannot be applied. This would require a PML boundary conditions in all directions.[1] For the realization of a plane wave, however, this problem does not occur. It is repeated periodically, similar to the unit cell of the PCS. Since we intend to perform transmission simulation under various angles, we have to include phase shift informations of the electromagnetic waves in the simulation domain and thus use BPBC for simulation walls in x and y directions (Fig. 3.2).

In the next step, the PCS itself should be modeled. In the lateral extension, the periodic arrangement of the PCS can be described in terms of its single unit cell. In Fig. 3.3

[1] In this case the cell size, in direction of the periodicity, needs to be large enough to approximate an infinity extended PCS.

the unit cells of the in this work frequently used PCSs are shown: linear, quadratic and triangular PCSs. These unit cells are described by the lattice parameters, which are the length of the cell edge (periodicity a) or the angle between grid points.

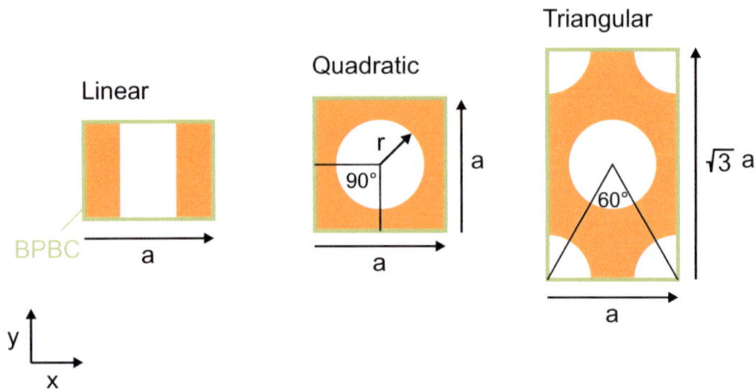

Figure 3.3: *Top view of the single unit cell of a linear, quadratic and triangular PCS. a is the periodicity of the structure, while r is the radius of the hole.*

In the direction normal to the surface (z-direction), the structure does not show any periodicity. Therefore we imposed PML boundary conditions at the top and bottom of the simulation domain.

As all PCSs used in this work were built on a substrate, the lower part of the simulation domain was filled with a glass or quartz material. It is inefficient to simulate the whole thickness of the substrate, as its thickness does not contribute to the guided modes in the PCS. Therefore, we terminated the simulation domain with a PML in the substrate material. The recorded electric and magnetic fields were normalized with respect to the refractive index of substrate.

Before setting the final simulation domain size in z direction, one more issue needs to be discussed: PML for evanescent waves. As discussed previously in section 2.2, photonic crystal modes, such as QGM, provide an evanescent wave outside of the structure. The wave vector k for such waves is complex, most commonly it is purely imaginary. A standard PML, however, can only handle positive and real k vectors. This problem is solved very trivially by choosing the computational region larger than the penetration depth of an evanescent wave, which can be up to half of its wavelength.

Besides the geometrical modeling of the structure, the modeling of material properties needs to be addressed as well. In general, materials are determined in Maxwell's equations by their relative permittivity ϵ and their relative permeability μ. Since the transmission simulation covers a wide range of frequencies the frequency dependency of these parameters must be considered as well (material dispersion). One way to describe these parameters is to use a sum of harmonic resonances plus a term from the frequency-independent electric conductivity. Materials, which can be described by this formalism are called *Lorentz* materials. The equation that describes the complex permittivity is the Lorentz equation:

$$\epsilon(\omega) = \epsilon_\infty + \sum_n \frac{\sigma_n \omega_n^2}{\omega_n^2 - \omega^2 + 2i\omega\gamma_n}. \tag{3.10}$$

Here, ϵ_∞ is the relative permittivity at infinite frequency, σ_n is the change in relative permittivity due to the Lorentz pole pair, ω_n is the frequency of the pole pair and γ_n is the damping coefficient. The Lorentz equation can only be applied to dielectric materials with losses. Modeling a metal at optical wavelength, however, the *Drude* equation is

$$\epsilon(\omega) = \epsilon_\infty - \sum_n \frac{\omega_n^2}{\omega^2 - i\omega\gamma_n}. \tag{3.11}$$

widely used.

The interested reader is referred to reference [91], [97] for a more detailed discussion of frequency dependent media.

3.2.2 Transmission in time domain and frequency domain

For reasons, which will be discussed in the next section, transmission through the PCS is recorded by the point monitor in the time domain. To obtain the transmission spectra as a function of the wavelength, the fast Fourier transformation (FFT) of the time domain data is used. The final transmission is obtained by normalization of this spectrum to the transmission spectrum of the source through vacuum, which is also obtained by Fourier transformation of the time domain transmission (Fig. 3.4).

At this stage we can discuss the position of the point monitor in the simulation domain. Because of the normalization of the transmission spectra to the spectra of the transmission through vacuum, an arbitrary position of the point monitor can be chosen. Again, we have the restriction to position the point monitor far away from the PCS, so that the QGM do not disturb the monitor. However, the point monitor should be placed in the center of the simulation domain, because the absolute field amplitudes have their maximum at this position.

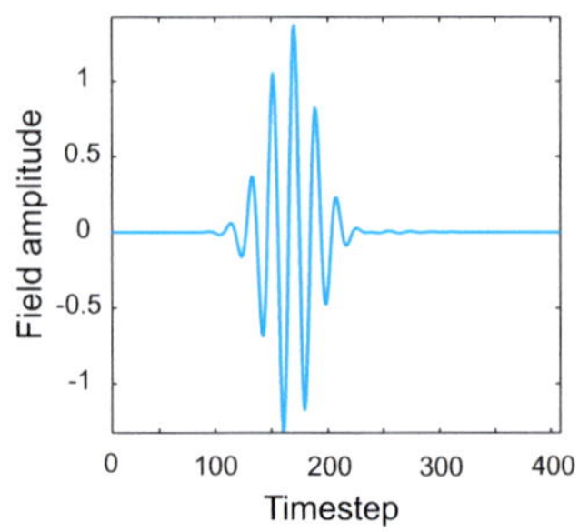

Figure 3.4: *Transmission through vacuum as a function of time step. The Gaussian pulse shape of the illumination source, which is a plane wave, is obtained.*

As an example, a quadratic PCS[2] with a high refractive index for the slab is chosen, this PCS shows pronounced photonic crystal features and is used throughout this chapter as the sample PCS. The simulation domain has an expansion of 400 nm in x and y direction, which defines the periodicity, and 2 μm in z direction. The plane wave Gaussian source was placed 800 nm above and the point monitor 500 nm under the center of the simulation domain.[3] As the illumination is under normal incidence, s [4] and p [5] polarizations yield the same results. In Fig. 3.5 the Fourier transform of the recorded field amplitude in time domain, which is the transmission spectrum at normal incidence through this PCS, is shown. As expected and discussed in section 2.4, GMRs, which show a strong Fano shape, are superimposed with a background transmission. The smoothly varying background is a consequence of Fabry-Perot oscillations due to the uniform dielectric slab. Although the dielectric slab shows a periodic pattern, the Fabry-Perot oscillations are not disturbed and only the effective refractive index of this slab is relevant for them.

Using this simulation approach, the response function of the structure can be obtained at a given wave vector k. Furthermore, using equation 2.10, a fano-fit can be applied to the spectra to obtain the line width of the resonances. Hence, we obtain the band structure of a PCS above the light line. It contains not only the spectral position but also the line width of each resonance. The s and the p polarizations are obtained by recording the

[2]The PCS is composed of a 200 nm silicon slab with a refractive index n = 3.46 and hole in a quadratic arrangement with a periodicity of 400 nm. The holes show a cylindrical shape and have a radius of 80nm.

[3]The PML boundary condition used 12 layers and the mesh was constructed by *FDTD Solutions* in a "auto non-uniform" mode.

[4]Polarization direction perpendicular to the plane of incidence. In German: **s**enkrecht.

[5]Polarization direction in the plane of incidence. In German: **p**arallel.

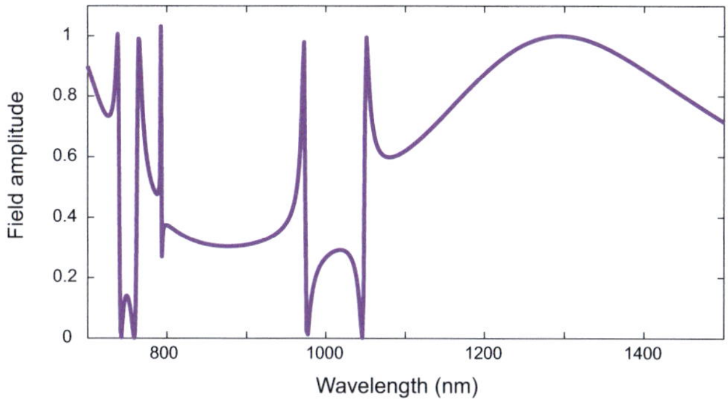

Figure 3.5: *Transmission through a quadratic PCS as a function of the wavelength. GMR are superimposed with a smoothly varying background.*

electric field in both directions, E_x and E_y respectively. For this the polarization direction of the illumination source should have preferably equal components in both directions.

In this work it will be shown that the direct access to the GMR will open a way to a variety of applications in biotechnology. In the experimental realization this is achieved by embedding the PCS between two crossed polarization filters. That way the broadband background light is suppressed and only GMRs are transmitted. This behavior needs to be modeled, which is the emphasis of the next section.

3.3 Direct access to guided-mode resonances

In order to have the direct access to the GMRs, a method based on FDTD, which is proposed by S. Fan and J. D. Joannopoulos in [10], is used in this work. The advantage of this method is the access to pure GMRs, independent of the polarization direction of the illumination source and the geometrical arrangement of the PCS. This method is based on the time domain analysis of the transmission through the PCS of a pulsed source. In Fig. 3.6 the transmission in time domain through the PCS, introduced in the previous section, is shown.[6] This data was recorded for a duration of 3000 fs. The most interesting part, which is the first 3000 time steps, is zoomed in and plotted in the lower part of this figure.

[6]These data were used to generate the transmission spectra in Fig. 3.5.

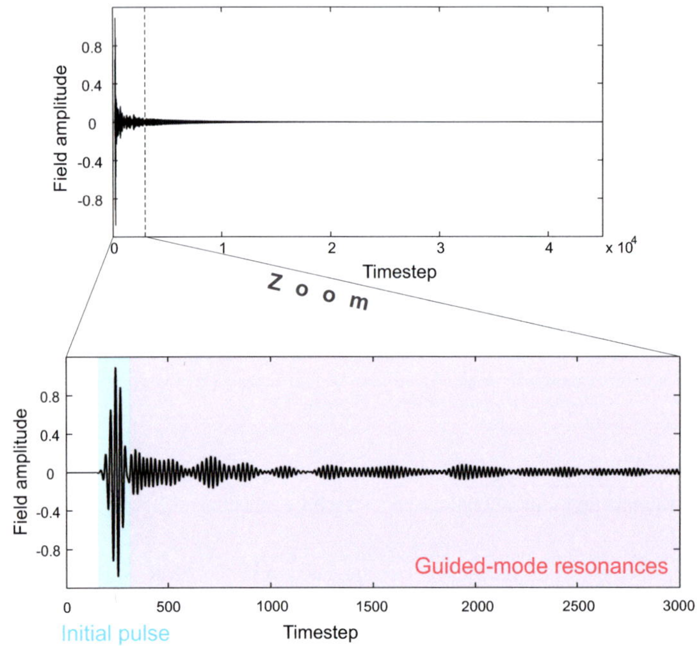

Figure 3.6: *Transmission through a quadratic PCS as a function of time step. The simulation duration of 50000 time steps, which corresponds in this case to 3000 fs, can be divided in two distinct time areas: the initial pulse (blue area), containing the direct transmission of the source and thin film effects and a tail of long delay (red area), which are the GMRs provided by the PCS.*

This time sequence can be divided into two stages, which are based on two different physical backgrounds. The first part (blue area) is an initial pulse, which is the direct transmission of the illumination source through the PCS. Comparing this part of the transmission with the transmission through vacuum, which is shown in Fig. 3.4, similarities can clearly be identified. In the first part, the interference information, which is the smoothly varying background, resulting from the thin slab, is also included. This initial pulse is followed by a tail of long decay (red area). After the illumination part has passed the simulation domain, the point monitor is still recording electromagnetic radiations. This electromagnetic radiation has its origin in the QGMs. Depending on the *quality (Q) factor* of each mode, the QGM is captured and at the same time propagating in the slab structure. During this propagation process it radiates its energy gradually. The core of this method is to separate these two time sequences and generate the transmission as a function of the wavelength using a Fourier transformation.

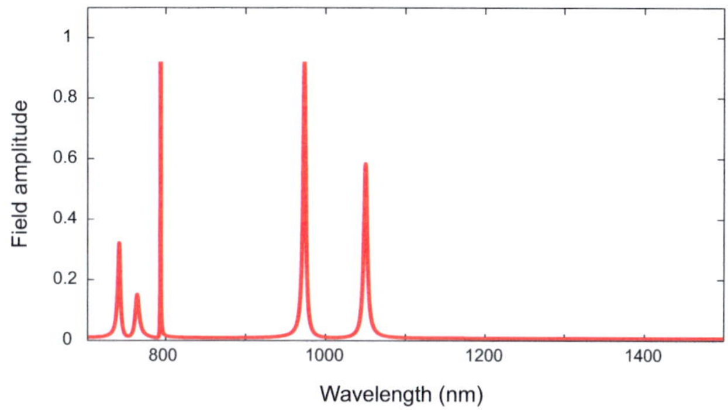

Figure 3.7: *Transmission through a quadratic PCS as a function of the wavelength after filtering. Only GMR are obtained.*

In Fig. 3.7 the Fourier transformation of the time sequence (as shown in Fig. 3.6) excluding the first 300 time steps is shown. Here, only the QGMs, which leave the structure with a delay depending on their Q-factor are visible. As discussed in section 2.4 these resonances are referred as to GMRs. At this point it should be underlined that also the line width of the GMRs can be obtained by this kind of simulations.

Reflection simulations can be performed in a similar manner, however details of these simulations are not discussed in this work. For more information the reader is referred to reference [10].

3.4 Impact of geometric parameters on guided-mode resonances

The geometric parameters of a PCS refers to the arrangement of the dielectric in space, i.e., hole radius and slab thickness as well as the hole shape. These parameters and the material properties, such as refractive index of the dielectric, affect the GMRs in terms of their spectral position or Q-factor. The physics behind this effect can be explained with various models. The spectral position shift of the GMR, when varying the periodicity

(Λ) of the structure or the effective refractive index (n_{eff}) of the slab material, can be modeled with the Bragg condition as introduced in equation 2.1. The Bragg condition is a good approximation in these cases, since QGM can be described as a guided mode in a slab structure with an effective refractive index n_{eff}, which is disturbed by a periodic scattering pattern. A lot of interesting phenomena occur, when the hole radius of a PCS is varied. This will be discussed in the next section.

3.4.1 Varying the hole radius

In Fig. 3.8 the transmission through our sample PCS as a function of the hole radius is shown. The hole is a cylinder filled with air and the radius is varied from zero to 200 nm, thus the maximum diameter equals the PCS periodicity. The most interesting phenomena regarding hole radius variation are GMR spectral position shift, Q-factor decrease with larger hole radius and GMR disappearance at large hole radii, which will be discussed in the following.

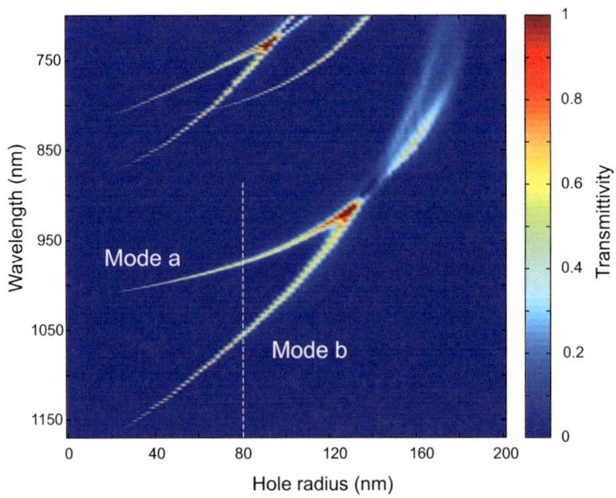

Figure 3.8: *Transmittivity through a quadratic PCS as a function of hole radius. GMRs experience a shift in spectral position and change in Q-factor. Furthermore no GMRs are observed for too small and too big hole radii.*

Maybe the most distinctive effect of varying the hole diameter is the resonance spectral position shift. We observed a shift towards lower wavelengths with increasing hole radius. This resonance shift is also explained by the Bragg condition. The n_{eff}, which is used in equation 2.1 is the effective refractive index of a mode, depending on the geometric arrangements of the PCS and field distribution of the mode. In our case, n_{eff} was decreasing, when the hole radius was increased. Therefore, GMRs were shifted towards lower wavelengths. Furthermore we observed different shifts per RIU (refractive index unit) of each mode. For example in Fig. 3.8 the resonance shift of mode a is less pronounced than the shift for mode b. To investigate this phenomenon in more detail, we plot the field distributions[7] of both modes in Fig. 3.9 at a hole radius of 80 nm.[8] This filed distribution reveals an antisymmetric field distribution of mode a, which is an odd mode and a symmetric field distribution for mode b, which is an even mode. Mode a, which has the smaller resonance shift, has its field minima and maxima on the surface of the PCS, whereas there is almost no field intensity in the center of the unit cell, where the hole is placed. Thus this modes's n_{eff} is not affected by hole radius variation. On the contrary, mode b has its field minima and maxima in the center and edges of the unit cell, thus it is highly disturbed by the hole radius variation. Calculating the n_{eff} for a mode requires the exact knowledge of its field distribution and the refractive indices of contributing materials. The following equation is used for this purpose:

$$n_{eff} = \sum_i a_i n_i. \tag{3.12}$$

Here, i represents different materials, such as slab or substrate material, n_i is the refractive index and a_i is the fraction of the mode penetrating the material i. The resulting n_{eff} is valid, as long as the field distribution of the mode does not change.
The reader may have noticed that in Fig. 3.9 the field distribution far away from the PCS is not dropping to zero. This is the radiation of the GMRs to the far field in both directions.

Investigating Fig. 3.8, we observe a lowering of the Q-factor with increasing the radius. Approaching smaller hole radii the QGMs become more strongly guided, which means that their lifetime increases. In the same manner, with growing radius, more scattering processes happen and the QGM radiates its energy faster to the far field. This observation can be used to design the Q-factor of GMR depending on the application.

Furthermore, we see from Fig. 3.8 that for a hole radius larger than 180 nm no modes are observed anymore. To explain this behavior, we argue again with n_{eff} of the mode, which decreases when the hole radius increases. Moreover we consider equation (2.9) on page 23, where a condition for guided modes is described. For a too low n_{eff}, the

[7] To obtain the field distribution we excite the PCS with a source with limited frequency range. Thus only one mode is excited and the overall field distribution is the field distribution of the mode of interest.

[8] Electrical field distributions in yz direction in the simulation domain at x=0.

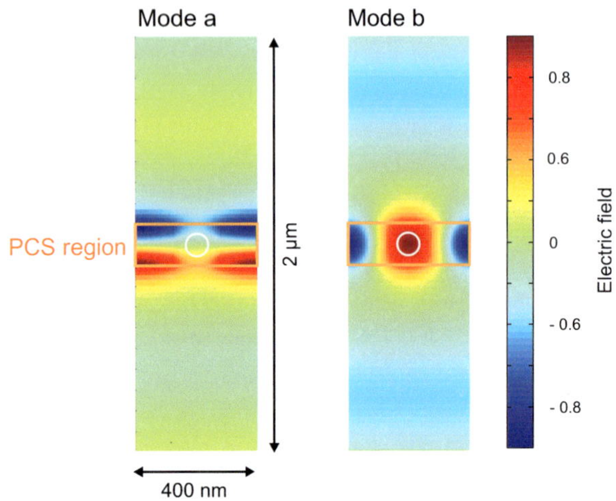

Figure 3.9: *Electric field distribution of mode a and mode b in the simulation domain in yz direction at x=0. Mode a has its field maxima at the surface of the PCS, while mode b has its maxima in the slab structure. This fact explains the different n_{eff}, which is proportional to the resonance shift.*

refractive index contrast to the cladding material, which is air in this case, cannot fulfill the condition in (2.9), hence no guided modes can occur.

Chapter 4

Transmission using crossed polarization filters

Summary

In this chapter we introduce a method for observing directly guided-mode resonances in photonic crystal slabs. It begins with description of the experimental setup of a confocal microscope for this method. Special focus is on developing a differential measuring method for angle resolved transmission measurements. This method is integrated in a confocal setup and thus shows also a high spatial resolution.
Further, in this chapter details and the physical background of applying crossed polarization filters for transmission measurements are discussed. This approach allows to obtain the guided-mode resonances, which are characteristic for the photonic crystals slab.

4.1 Experimental setup

Before discussing the transmission method using crossed polarization filters, the experimental setup is introduced. In the following a setup is described, which not only allows for spatially resolved, but also for angle resolved transmission measurements. A commercially available inverse microscope[1] was modified with a pinhole to obtain a confocal setup, which allows for spatially resolved measurements. Moreover, a stepper motor was applied to vary the the size of the illumination's iris diaphragm to control the angular resolution.

4.1.1 Confocal microscope setup for spatially resolved transmission measurements

The experimental setup is shown schematically in Fig. 4.1. The sample is placed on a motorized translation stage[2] in a confocal microscope setup. This translation stage offers a resolution down to 500 nm and could be extended with a rotation stage for the azimuthal angle ϕ when needed. A 100 W halogen lamp was used as a broadband illumination source. For a homogeneous illumination, the bulb surface of the lamp was roughened. Two orthogonally oriented polarization filters (extinction ratio 8,500:1), which were both rotatable in their azimuthal angle, were placed before and behind the photonic crystal slab. We refer to the first polarization filter as *polarizer* and to the second polarization filter as *analyzer*. The polarizer is placed between the halogen lamp and the condenser, and the analyzer behind the objective. We could choose between four magnifications of the objective: 4x with a numerical aperture of 0.1, 20x with a numerical aperture of 0.4, 40x with a numerical aperture of 0.6 and 100x with a numerical aperture of 1.4. Moreover, in the 20x objective a phase ring is integrated that in combination with a matching annular ring in the condenser allows for phase contrast microscopy. We used a long distance condenser with a working distance of 75 mm. A slit-diaphragm defines the polar illumination angle θ.

The crucial part of the confocal setup is a 100 μm pinhole, which is placed in the focal plane of one of the microscope output ports. In combination with the 40x objective, for instance, the spatial resolution achieved by this system, is 2.5 μm. Further, two lenses are used to couple the light passing through the pinhole into a spectrometer. This spectrometer is composed of an imaging monochromator[3] and a back illuminated CCD camera[4] with a

[1]TE 2000 *Nikon*
[2]Corvus *Märzhäuser*
[3]Triax 320 *HORIBA Jobin Yvon*
[4]Symphony *HORIBA Jobin Yvon*

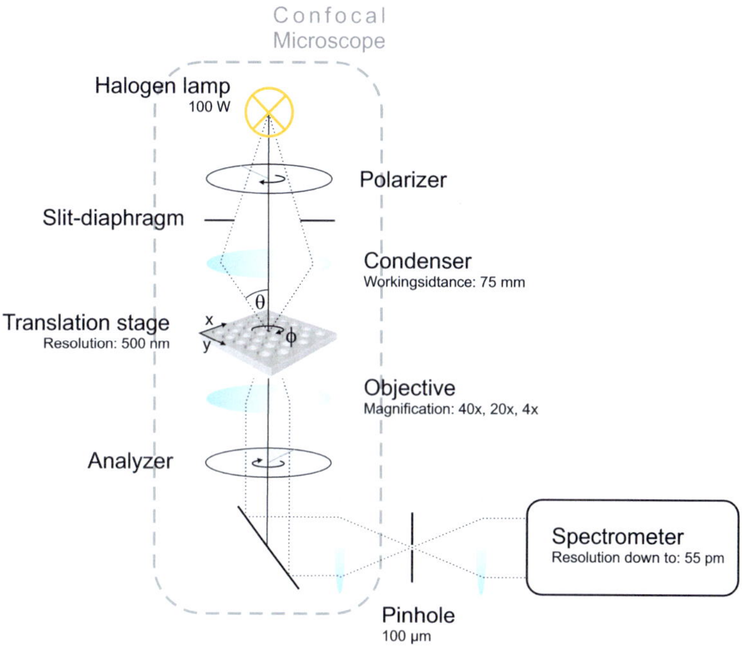

Figure 4.1: *Confocal microscope setup for transmission measurements with orthogonally oriented polarization filters. The first and second polarization filters are referred to as polarizer and analyzer. ϕ and θ are azimuthal and polar angle respectively.*

maximal spectral resolution of 55 pm.

4.1.2 A differential method for angle resolved transmission measurements

For characterizing PCS in the reciprocal space, angle resolved measurements are essential. Performing angle resolved transmission measurements, while keeping the spatial resolution, however, is an ambitious task. Maintaining this spatial resolution is an important requirement of our setup, specially for inhomogeneous PCSs, where parameters such as spectral position or linewidth of GMRs, vary with position. Therefore, a wide area spectrum would yield a blurring of these parameters.

To solve this task, we used a built-in iris diaphragm, which is placed above the condenser. This iris diaphragm limits the numerical aperture of the condenser, which focuses the light of the halogen lamp on a defined point on the sample. Thus, a light cone with the angle θ will illuminate the sample. The objective collects the transmission and directs it to the spectrometer. The numerical aperture of the objective should be higher or equal to the numerical aperture of the condenser, to be able to collect the desired angle. Since the maximum numerical aperture of the condenser is 0.3, both objectives with 20x and 40x magnification can be used. However, transmission measurements with this cone contain a continuum of light at angles from zero to θ. For band structure measurements we need the transmission of a discrete angle.

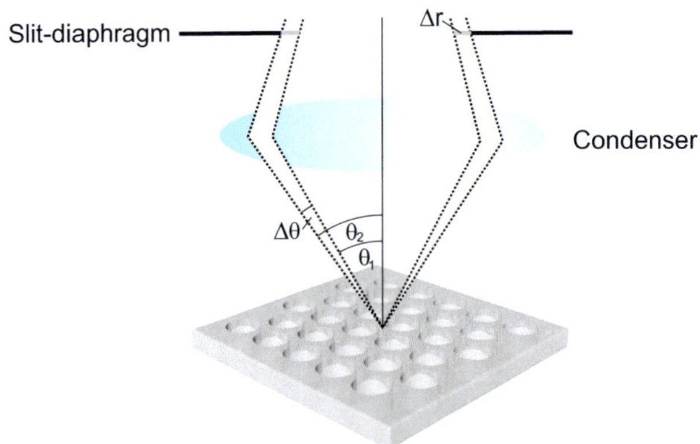

Figure 4.2: *Sketch of the differential method for angle resolved transmission measurements. By opening the slit-diaphragm incrementally and subtracting consecutive results, this differential measurement for angular illumination is obtained.*

To obtain transmission data for discrete angles, a differential method is used. The iris diaphragm is opened incrementally and a series of transmission spectra are recorded. The transmission at a discrete angle θ is generated by subtracting consecutive results as shown in Fig. 4.2. With this method, however, the effective illumination still exhibits a small angle continuum $\Delta\theta$, which originate from the discrete step size Δr. For our measurements the iris diaphragm was motorized and a software[5] was developed to control the step motor and the spectrometer.

Considering the band structure of a PCS, such as the triangular example in Fig. 2.10, one can observe a different progress of modes for different directions, for instance for

[5]Labview *National Instruments*

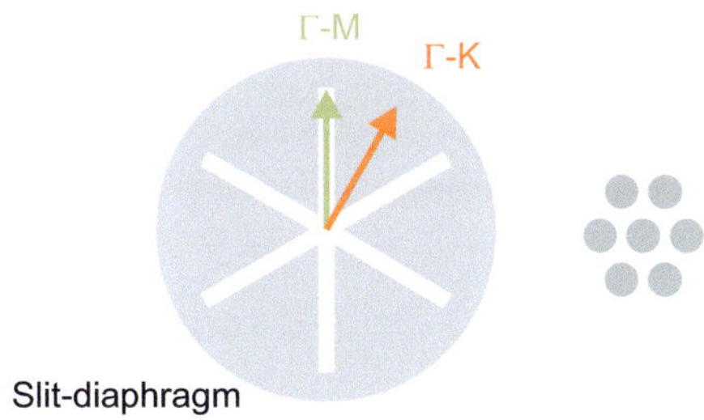

Figure 4.3: *Top view of slit-diaphragm and its orientation to a triangular PCS. A rotation of 30°, 90°, 150° etc. will switch from* Γ *- M to* Γ *- K direction.*

$\Gamma - M$ and $\Gamma - K$ direction. Using only the iris diaphragm, transmission spectra would contain all directions and as a result the modes would be blurred. This problem is solved by applying a shadow mask placed near the iris diaphragm. We refer to the combination of the iris diaphragm and the shadow mask, as *slit-diaphragm*. This shadow mask offers a slit showing in the direction of interest. Depending on the periodicity class of the PCS under investigation, this slit can be repeated at defined azimuthal angles to enhance the illumination. As an example for a triangular PCS, the slit can be repeated at every 60° as shown in Fig. 4.3. To switch from Γ-M to Γ-K direction, in the case of the triangular PCS, a rotation of 30°, 90°, 150° and so forth is applied.

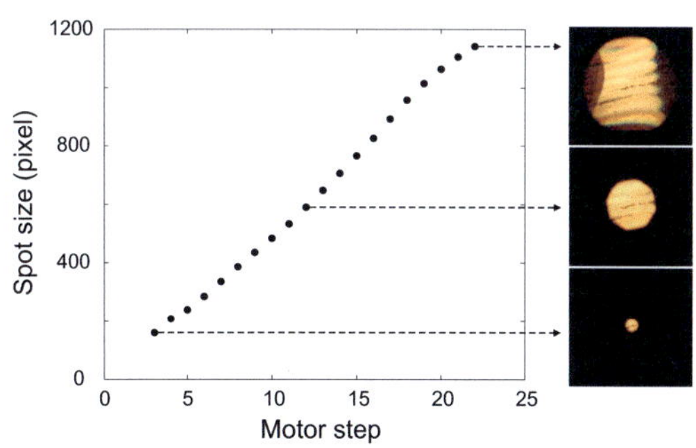

Figure 4.4: *Calibration curve of the angle θ versus steps of the step motor. A linear relation for the spot size, which is used to calculate angle θ, to steps of the step motor is observed. On the right hand side camera images of the illumination source at different opening angles are shown. Here the bulb surface is not roughened as used later in experiments and the shadow mask is removed for a better evaluation of the spot size.*

Since we use an external step motor to adjust the size of the iris diaphragm, a calibration of the system was performed. For this purpose we placed a camera between the condenser and the sample to monitor the size of the illumination spot and plot the step size as a function of the spot size. The calibration curve is shown in Fig. 4.4 and exhibits a linear relation between the spot size and steps of the step motor. From this curve we calculate a minimum angle[6] of 4° and a maximum angle[7] of 17° for θ.

4.2 Direct observation of guided-mode resonances

In previous work observing GMRs for sensor applications or PCS characterization [62,98] was performed indirectly by using conventional transmission or reflection measurements. The direct access to these modes opens up new design possibilities for sensor applications and simplifies characterization methods.

In this work a spectroscopy method is introduced, with which transmission measurements are performed with two crossed polarization filters placed before and behind the PCS. This configuration suppresses the light not interacting with the PCS and allows the direct observation of the GMRs with their Lorentzian line shapes [99].

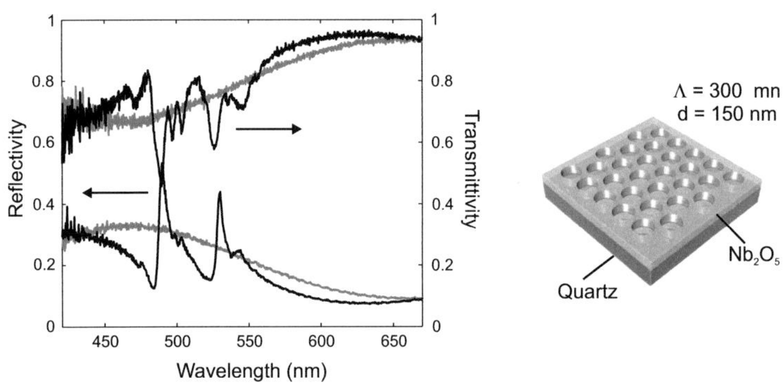

Figure 4.5: *Transmittivity and reflectivity measurements of a PCS with triangular geometry using parallel polarization filters. Black lines indicate measurements on structured and gray lines on unstructured area at (θ = 0°). On unstructured area thin film interferences cause a smoothly varying transmission. On structured region GMRs are superimposed with smoothly varying background.*

To demonstrate this effect and as a first example, the characterization of a photonic crystal slab, which exhibits air holes (r = 65 nm) in a Nb_2O_5 layer (d = 150 nm) on a quartz sub-

[6]Limited by the smallest size of the iris diaphragm.
[7]Limited by the numerical aperture of the condenser.

strate with a periodicity of $\Lambda = 300$ nm is discussed (PCS_{Nb2O5} in table 2.2). The structure was fabricated by electron beam lithography and a subsequent dry etching process. The fabrication procedure for this structure and all other electron beam lithography samples discussed in this work is detailed in [100,101]. Fig. 4.5 shows transmittivity and reflectivity measurements at normal incidence for parallel orientation of polarizer and analyzer on structured (black line) and unstructured (gray line) surface areas. The transmitted intensity is normalized to the detected intensity without a sample for a parallel orientation of polarizer and analyzer. The reflected intensity is normalized to the reflection on a quartz surface.

In the unstructured area, Fabry-Perot oscillations for both transmission and reflection are due to the thin-film effects clearly visible. In the structured area, however, these thin-film effects are superimposed with photonic crystal GMRs. As discussed in [10], these resonances are referred to as Fano-like resonances.

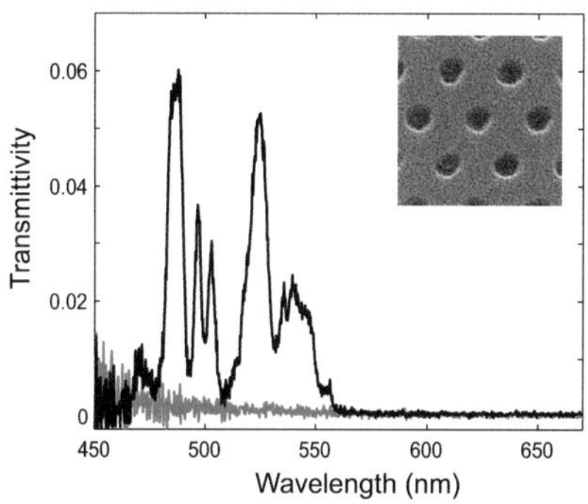

Figure 4.6: *Transmittivity measurements of a PCS with crossed polarization filters. Here only light interacting with the PCS can pass the analyzer, hence only GMR are transmitted. These resonances show as expected a Lorentzian line shape. Inset: SEM image of triangular PCS.*

If the same measurement is repeated in transmission with crossed polarization filters, the smoothly varying background is suppressed (Fig. 4.6). The background light that does not interact with the PCS, is blocked by the analyzer. Light from the light source, which can excite QGMs, is coupled to the PCS and coupled out again. During this coupling process there is a polarization rotation, which allows the transmission of the GMRs through the analyzer. This mechanism results in a pure transmission of GMRs. As expected these resonances exhibit a symmetric Lorentzian line shape [10].

4.3 Impact of polarization direction

To investigate the effect introduced in the previous section and understand the polarization rotation mechanism, in the following more simple PCS geometries, such as linear or quadratic PCS, are considered. Furthermore the impact of the polarization direction on the coupling mechanism is discussed.

4.3.1 Coupling mechanism to quasi guided modes

The unpolarized illumination source can be divided into two polarization directions: s and p. s-polarization is defined for electromagnetic waves with the electric field perpendicular to the plane of incidence. On the other hand, p-polarization is defined for waves with electric field parallel to the plane of incidence. In the following, we consider a linear PCS to simplify the coupling mechanism. As defined in Fig. 4.7, illuminating a PCS at $\phi = 0°$ s-polarization can excite only odd modes,[8] with electric fields parallel to the PCS's grooves. Contrary to s polarization the p polarization can excite only even modes. However, for $\phi \neq 0°$ s-polarized electromagnetic waves do not show exclusively in the direction parallel to these grooves. Therefore, s-polarization can also excite even modes, following the projection of polarization direction onto the direction of the groove.[9] Obviously, for $\phi = 45°$ both, even and odd modes, can be excited equally, either using s or p polarization.
As discussed in section 2.4, these modes leak out to the far field. The polarization for even modes is in the direction of the periodicity and for odd modes shows perpendicular to the direction of the periodicity.

If a polarization filter (analyzer) is placed after the PCS, as depicted in Fig. 4.8, and $\phi \neq 0°$, both modes can pass the polarization filter. Here, the intensities are calculated from the projection of the polarization direction onto the direction of the polarization filter. Now combining both, the excitation with polarized light using the polarizer and the detection with the analyzer, it can be shown that for orthogonally oriented polarization filters, the excited modes can pass the analyzer. Simultaneously, the light not interacting with the PCS is blocked by the crossed configuration of polarizer and analyzer [102].

4.3.2 Linear photonic crystal slabs

To verify these assumptions first the most simple case, i.e., is the transmission of GMRs of a linear PCS as a function of the azimuthal angle ϕ, is studied. These PCSs are

[8]Only leaky modes above the light line.

[9]In a similar way p polarized light excite both, even and odd modes.

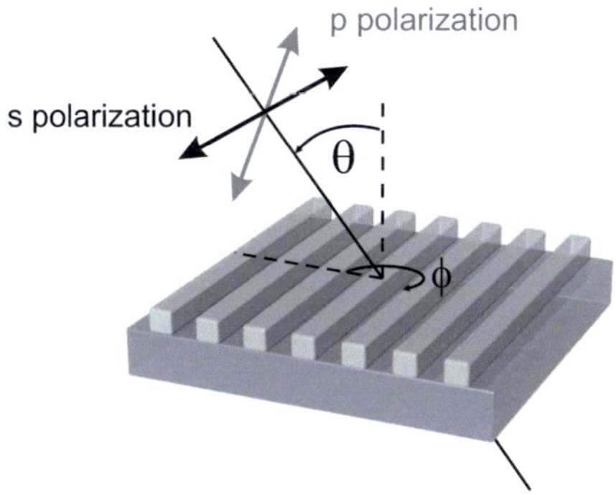

Figure 4.7: *Definition of the illumination azimuthal angle ϕ and polar angle θ. For a linear PCS $\phi = 0°$ for grating grooves perpendicular to the plane of incidence. $\theta = 0°$ for illumination at normal incidence. For s polarization the electric field is perpendicular and for p polarization parallel to the plane of incidence.*

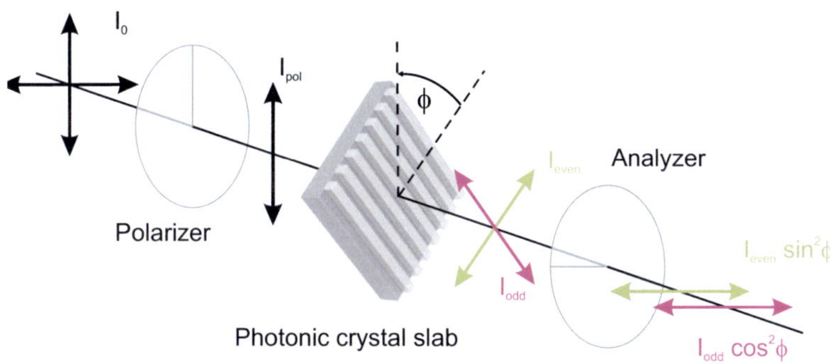

Figure 4.8: *Schematic of photonic crystal orientation relative to polarizer and analyzer for normal incidence measurements. Arrows indicate electric field orientation. Azimuthal angle ϕ defines the fraction, which couples to the photonic crystal slab.*

fabricated using a combination of laser interference lithography and physical plasma etching and consist of a 135nm ITO film, which has a linear grating with a periodicity of 350nm (PCS$_{ITO}$ tale 2.2). The inset of Fig. 4.9 shows a scanning electron microscope's (SEM) top view image of the used PCS.

In Fig. 4.9 transmission measurements with crossed polarization filters are shown. The broadband background light is suppressed and only guided-mode resonances are transmitted. There is no transmission of guided-mode resonances at $\phi = 0°$, while at other angles transmittied light is observed ($\phi = 20°, 40°$). This relation is shown more precisely in Fig. 4.10, where angle ϕ is varied from $0°$ to $180°$ in $5°$ steps and the intensity of the central mode (530nm - 540nm) is plotted. This transmittivity exhibits its maxima at $\phi = 45°$ and $135°$ and decrease to zero for $\phi = 0°, 90°$ and $180°$.

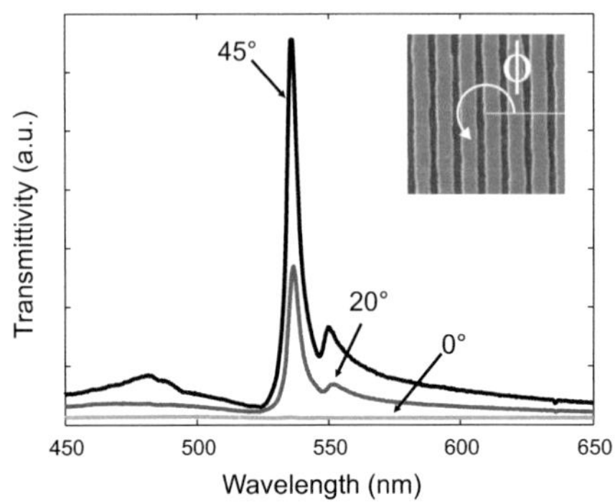

Figure 4.9: *Transmission measurements with crossed polarization filters on a linear PCS ($\theta = 0°$) at three different angles ϕ. For $\phi = 0°$ no transmission is observed. Inset: SEM image of linear PCS.*

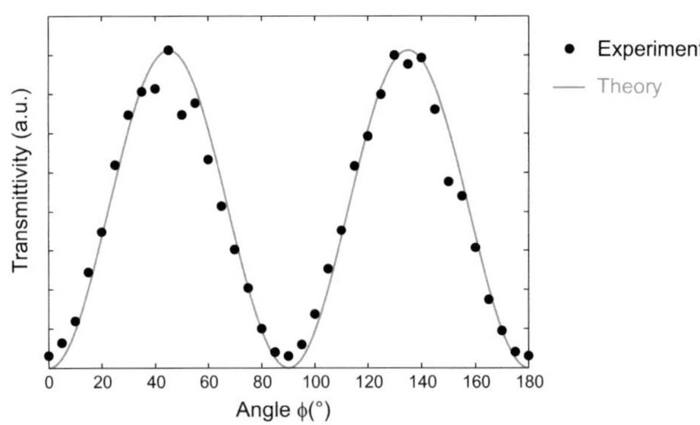

Figure 4.10: *Transmission of central mode (530nm - 540nm) intensity versus angle ϕ. Dots are measured data, while the solid line is the analytical model derived from Malus' law. Maxima are observed at $\phi = 45°$ and $135°$.*

In Fig. 4.11 the assumptions of the previous section for a linear PCS are used to calculate the angular dependence of mode intensity for even modes. I_{pol}, which is the intensity of the illumination source after the polarizer, is $1/2I_0$ following Malus' law, where I_0 is the intensity of the unpolarized light. Again, applying Malus' law, the fraction of I_{pol}, which couples to the PCS, is $I_{even} = I_{pol}\cos^2\phi$. Furthermore, the fraction of the mode, which is radiated to far field and can pass the analyzer, is $I^*_{even} = I_{even}\sin^2\phi$. From this series of transmissions through a polarization filter and coupling and out coupling from the PCS, the following equation can be derived for the intensity dependence of the GMR:

$$I_{odd}(\phi) = I_{even}(\phi) = \frac{1}{2} I_0 \cos^2\phi \ \sin^2\phi. \tag{4.1}$$

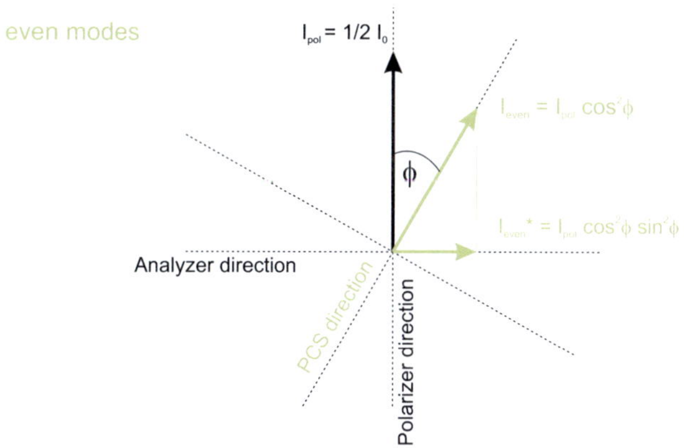

Figure 4.11: *Projection of polarizer and analyzer onto a linear PCS. Using Malus' law the intensity of the transmission of GMR (here for even modes) versus angle ϕ is modeled.*

The solid line in Fig. 4.10 is obtained by using equation 4.1.

4.3.3 Quadratic photonic crystal slabs

To have a closer look into the polarization rotation process, a higher symmetry PCS, namely a quadratic PCS, is investigated. The PCS used in this section is fabricated using the same technology and has the same periodicity and slab thickness as the one used in the last section. In order to fabricate the quadratic geometry the samples are rotated by $90°$ degrees during the lithography process and exposed a second time. An SEM image of the used samples is shown in the inset of Fig. 4.12.

In the following, a simplified picture of a quadratic PCS is used, where two orthogonally oriented linear PCS on one substrate, represent the quadratic geometry. Applying the conclusions of the previous sections and using the simplified picture of the quadratic PCS, the sum of $\frac{1}{4}I_0\cos^2\phi\ \sin^2\phi$ for the first linear PCS and $\frac{1}{4}I_0\sin^2\phi\ \cos^2\phi$ for the second one is the transmission of GMR. This results again in equation 4.1, which is the transmission for a linear PCS.

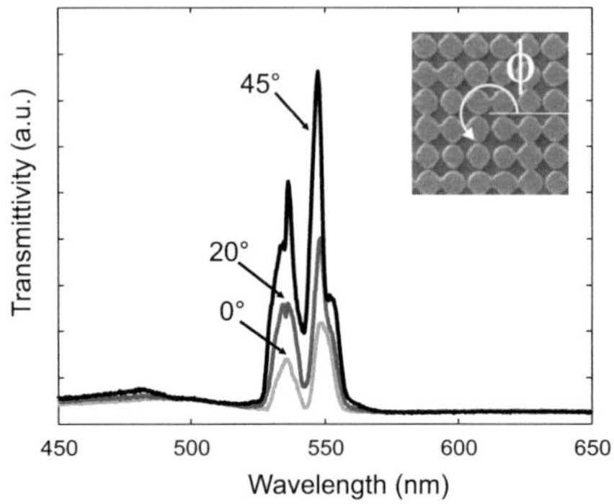

Figure 4.12: *Transmission measurements with crossed polarization filters on a quadratic PCS ($\theta = 0°$) at three different angles ϕ. In contrast to the transmission through a linear PCS, transmission unequal to zero for $\phi = 0°$ is observed. Inset: SEM image of quadratic PCS.*

Transmission measurements, however, show another result. In Fig 4.12, again the transmission with crossed polarization filters at normal incidence ($\theta = 0°$) is shown. Contrary to expectations, the GMR transmitted at $\phi = 0°$ is not zero. This behavior can be better observed in Fig. 4.13 where the intensity of the first mode (530nm - 540nm) as a function of angle ϕ is plotted. Once again we see maxima at $\phi = 45°$ and $135°$, but the intensity of the GMR does not decrease to zero at the minima ($\phi = 0°, 90°$).

Form this experiment we concluded that the assumption that a quadratic PCS is the superposition of two orthogonally oriented linear PCSs, is not sufficient. To explain the observed effect in Fig. 4.13, it is necessary that mode coupling inside the PCS is considered as well. Modes, which are excited in the first linear PCS couple to modes in the second linear PCS and vise versa. We assumed that the coupling quotient for the first PCS to the second PCS is the same as the second PCS to the first PCS. Using a as the coupling factor and the projecting polarizer and analyzer onto the PCS, as shown in Fig. 4.14, we can derive for the intensity of modes coupling from the first PCS to the second one $a\cos^2\phi\ \cos^2\phi$ and vise versa from the second PCS to the first PCS $a\sin^2\phi\ \sin^2\phi$. Summing up these results with the fraction of the modes not involved in coupling processes, we end up in the following equation for the angle dependent GMR transmission:

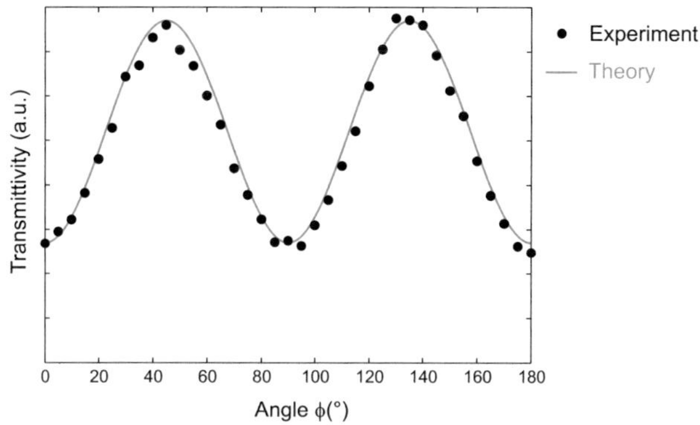

Figure 4.13: *Transmission of first mode (530nm - 540nm) intensity versus angle ϕ. Dots are measured data, while the solid line is the analytical model derived from Malus' law. Maxima are observed at $\phi = 45°$ and $135°$, whereas the minima don't decrease to zero.*

$$I_{odd}(\phi) = I_{even}(\phi) = \frac{1}{2} I_0 \left[(1-2a)\cos^2\phi\ \sin^2\phi + a\sin^2\phi\ \sin^2\phi + a\cos^2\phi\ \cos^2\phi\right] \tag{4.2}$$

The solid line shown in Fig. 4.13 is the best fit of equation 4.2 to the measured data, which resulted in a coupling factor of $a = 9\%$ for the first mode. For the second mode, shown in Fig. 4.12 (540 nm - 560 nm), the fitting procedure resulted in a coupling factor of $a_2 = 15\%$ (results not shown here).

These considerations can be also applied for higher symmetry PCSs such as triangular PCSs. In a triangular PCS a superposition of three linear PCSs with mode coupling is assumed. This results in an angle dependent GMR transmission, which is less pronounced than for a linear or quadratic PCS.

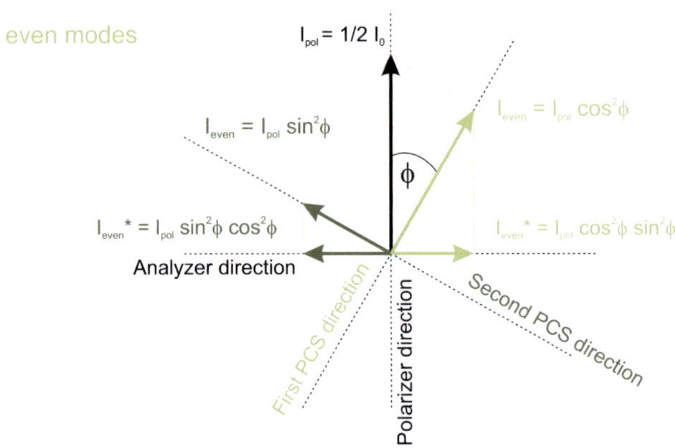

Figure 4.14: *Projection of polarizer and analyzer onto two orthogonally oriented linear PCSs. Using Malus' law the intensity of the transmission of GMR (here for even modes) versus angle ϕ is modeled. For each mode individually a coupling between modes excited in the first PCS and second PCS is assumed.*

Chapter 5

Methods for characterizing photonic crystal slabs

Summary

As photonic crystals and in particular PCSs are expected to be produced in future on an industrial scale, quality control is a crucial point. In this chapter the concept of direct access to the GMR is applied to characterize PCS properties.

PCS parameters such as hole radius or hole shape affect significantly the optical properties of such structures. In the first part of this chapter these parameters are obtained using parameter deduction, as SEM images can contain imaging aberration. Further in this chapter homogeneity investigations are performed, which unravel fabrication imperfections.

Another emphasis of this chapter is to characterize GMRs. Beside the angle resolved spectral position, the Q-factor of GMRs is investigated. Further the Q-factor determination and homogeneity investigations are combined and a Q-factor map of the PCS is generated.

5.1 Motivation

PCSs are promising structures for modern optical devices. The fabrication of such structures is realized employing different methods, for example, electron beam lithography or laser interference lithography. Each fabrication method, however, shows characteristic advantages and disadvantages. Hence, a proper quality monitoring is required, which visualizes imperfections. For proper quality monitoring, imaging methods such as SEM or atomic force microscopy (AFM) are not sufficient, since they only provide geometric parameters such as the periodicity or the hole radius. The exact knowledge of GMR spectral position and Q-factor in the reciprocal space is as important as information on the spatial homogeneity of these parameters across the PCS.

In previous investigations the spectral position and the linewidths of GMRs have been studied in the reciprocal space [62, 98] using conventional transmission measurements. As discussed in section 2.4 and theoretically in section 3.2.2 the transmission spectrum is a superposition of the smoothly varying background light superimposed with GMRs, which show a Fano shape [103]. To obtain GMR parameters such as Q-factor, an Airy-Fano model is necessary for the fitting procedure. However, the used methods so far provided a lower limit on the resolvable linewidth and hence showed disagreements comparing theory with experimental results. Another method is to employ the luminescence spectrum of a proper emitting layer inside the PCS, which reveals the GMRs, with its Lorentzian line shape [104–107]. From these spectra the Q-factor can be determined by applying the corresponding fit procedure. In passive structures, however, this method cannot be applied.

On the other hand many theoretical models are known for calculating spectral position and losses of GMRs [10, 55, 108, 109] that can be used to calculate GMR parameters. For real PCSs, however, these parameter are strongly affected by geometric parameters not only by the periodicity or hole size, but also parameters such as hole depth and shape [110]. Therefore, the experimental characterization is an essential procedure.

In this work an experimental method for a rapid and non-destructive local optical characterization of photonic crystal slabs is proposed. Transmission measurements are performed on photonic crystals slabs employing the crossed-polarizer configuration in the confocal microscope setup. This allows for the direct characterization of the GMRs of microscopic structures. In combination with the translation stage the setup is utilized to characterize the spatial homogeneity of macroscopic nanostructured surfaces. Unlike lateral transmission measurements these measurements are independent of the lateral termination of the structure and no special preparation is required [111].

5.2 Deduction of photonic crystal slab parameters

As already discussed in section 3.4 geometric parameters of PCSs influence significantly the GMRs. Therefore the exact knowledge of these parameters is crucial. Imaging methods, such as SEM imaging, can be used to read out these parameters from the images of such a structure. Using this approach, however, only approximate parameter values are obtained. First, aberration of these systems needs to be considered, which is not necessarily a trivial issue. Second, the surface of different materials shows different roughnesses, which complicates the determination of geometrical parameters.
Using an optical approach, which probes directly the optical properties of PCS and deduces the geometrical parameters, is a straightforward concept, as in general the optical properties of a PCS are the one of interest.

In the following we determine the material and geometrical parameter of a PCS consisting of a Nb_2O_5 layer with triangular lattice of air holes. This structure is chosen for the parameter deduction, as it was used for different investigations in the scope of this work.

5.2.1 Determination of slab thickness

Before treating the PCS itself, the unstructured slab is considered. The Nb_2O_5 slab structures, which are sputtered on a quartz substrate, are delivered by an external manufacturer. Unfortunately, the manufacturer's specification of the slab thickness possess high tolerances. Therefore, this specifications could just be used as a starting point for more precise thickness determinations.
There exist several methods to determine the slab thickness of a dielectric, such as ellipsometry or using a profilometer to determine the thickness on an edge of the slab. In this work transmission measurements were used to determine the slab thickness accurately. Due to thin film interferences, the transmission through an optical thin film is unique and is a function of its thickness. This effect is utilized to determine the slab thickness. The transmission through the slab was simulated using the FDTD method and compared to experimental transmission measurements of the slab of interest. Through variation of the slab thickness the difference between simulated and measured data is minimized and the best fit can be found.

To determine the slab thickness using this optical method, first material parameters have to be addressed. Besides geometrical input, material properties have to be defined in the simulation environment. As introduced in section 3.2.1 a common model for dispersive dielectric is the Lorentz model. In our case, only the high index layer (Nb_2O_5) is considered as dispersive and Lorentz parameters are determined only for this material. The real and imaginary part of the refractive index[1] is used to fit a Lorentz equation with one pole. The experimental data (dots) with the corresponding fit (solid line) is shown in Fig. 5.1. The Lorentz parameters extracted from this fit are summarized in table 5.1.

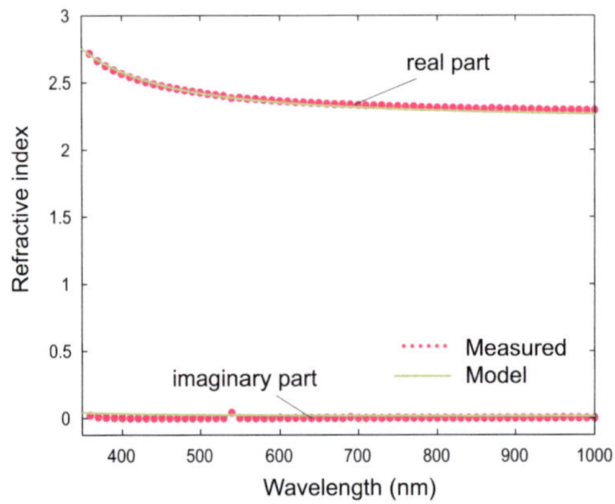

Figure 5.1: *Real and imaginary part of the refractive index of Nb_2O_5 with the corresponding Lorentz model. The employed Lorentz model exhibits one pole, which does not lies in the range of interest.*

ϵ_∞	1.74
σ_0	3.2
ω_0	8000 Thz
γ_0	-2π

Table 5.1: *Lorentz parameters for Nb_2O_5 as defined in section 3.2.1.*

These parameters were used to calculate the transmission through the slab as described in chapter 3. The best agreement of the simulated transmission spectra and measured spectra was obtained for a slab thickness of 221 nm, which is plotted in Fig. 5.2.

5.2.2 Adaption of hole radius and profile

Using the results of the previous section the periodic nanostructure can be modeled. Due to the fabrication method, in particular the lithography step, the periodicity of the

[1]These experimental data were obtained from the database of an ellipsometer.

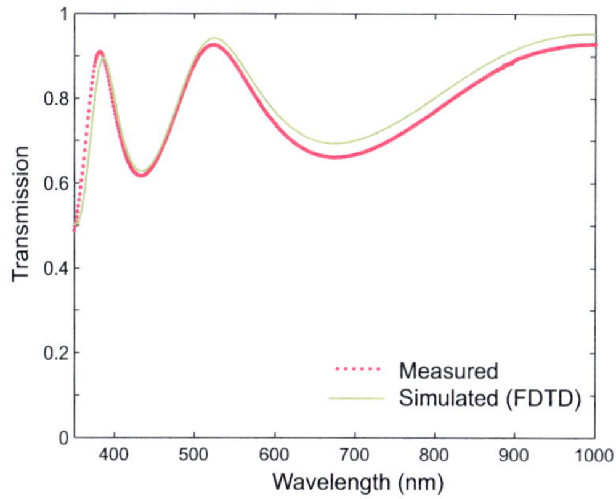

Figure 5.2: *Transmission spectra through an unstructured Nb_2O_5 layer: Experimental data (dots) and FDTD simulation (solid line). Due to reflections on the layer interfaces interferences are observed, which are the origin of the varying intensity as a function of wavelength.*

structure can be adjusted. Using electron beam lithography, for example, the beam position is adjusted by a pair of deflector plates and is at least as precise as the spatial resolution of a scanning electron beam microscope.[2] The exposure dose and the following etching process, however, do not allow always a predictable hole size and profile. Hence, many of fabricated PCS for this work, show a non cylindrical hole profile. As seen in the scope of this work and in the literature [110] the hole profile has a significant influence on the GMR spectral position and Q-factor.

In a first step a top view SEM images is used to estimate the hole radius (Fig. 5.3 left image). To have access to the hole profile, which cannot be revealed using only SEM, a focused ion beam (FIB) is applied to cut vertically into the PCS and open up the view on the cross section of the PCS, which is imaged by SEM at the same time. FIB is a technique, which uses a focused beam of gallium ions (Ga^+) for analysis, deposition, or ablation of materials. The cross section of the PCS is shown in Fig. 5.3 right image. This picture bares two main aspects, which are implemented into the simulations. First, the hole profile, which follows a polynomial shape due to the imperfect etching process. It showed up that a fifth degree polynomial describes best this shape. Second the holes are not etched all the way to the substrate and hence show an offset of 32 nm.[3]

[2]Depending on the instrument, the resolution can be less than 1 nm.

[3]This parameter was read out from the SEM image as no other method was known, which delivers this

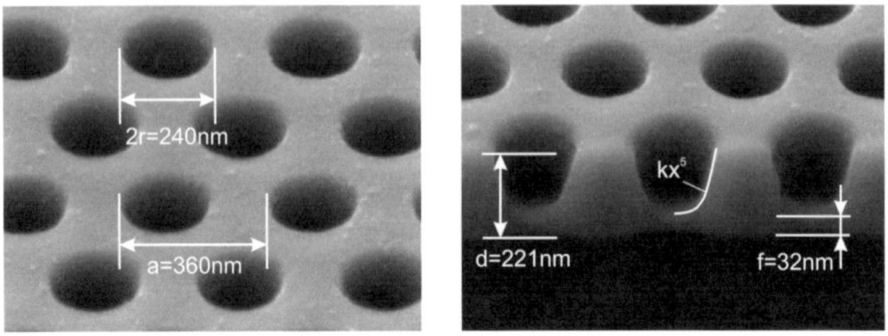

Figure 5.3: *SEM image of a PCS composed of a Nb_2O_5 layer with triangular lattice of air holes. Left image: Top view of PCS. Periodicity ($a = 360nm$) is adjusted by the electron beam lithography system. The hole radius ($2r = 240nm$) is obtained by deduction, by comparing simulations with experimental transmission measurements. Right image: SEM image of the cross section obtained by FIB. The hole profile can be described by a polynomial of degree 5. The slab thickness ($d = 221nm$) was obtained by investigating an Nb_2O_5 thin film. The offset of the hole relative to the substrate ($f = 32nm$) was read out from the SEM image.*

With this knowledge about the PCS a series of transmission simulations, varying the hole radius, was performed. As a starting point the hole radius obtained by SEM image analysis was applied. In Fig. 5.4 three calculated transmission spectra with different hole radii at the polar angle of $\theta = 5°$ are shown. This polar angle is chosen as it exhibits a richer spectrum of GMRs and thus allows for more accurate comparison to the experimental results than $\theta = 0°$. As expected bigger hole radii cause a decrease of the effective refractive index (n_{eff}) of the slab material and hence a resonance shift towards smaller wavelengths.

The results in Fig. 5.4 are compared to experimental measurements at $\theta = 5°$ of PCS of interest. As depicted in Fig. 5.5 the best fit is obtained for a hole radius of 120 nm.

5.3 Homogeneity characterization of photonic crystal slabs

Many applications of PCSs require large area structures, such as photonic crystal band pass filters or PCSs as transducers in biosensors [81] and in particular for applications

parameter in a reasonable time scale.

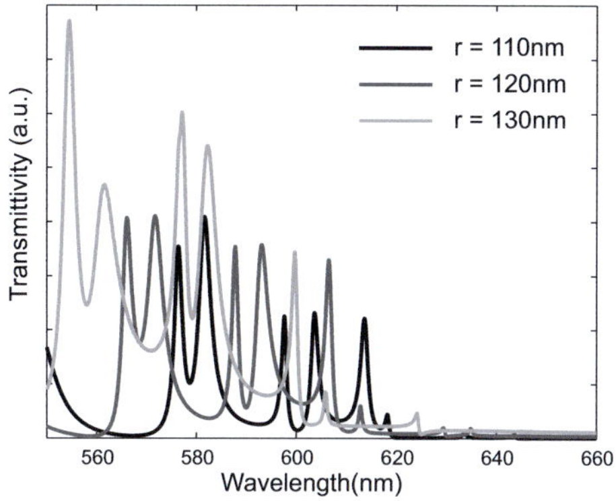

Figure 5.4: *FDTD transmission simulations with varying hole radius at the polar angle of $\theta = 5°$. Bigger hole radii cause a resonance shift towards smaller wavelength.*

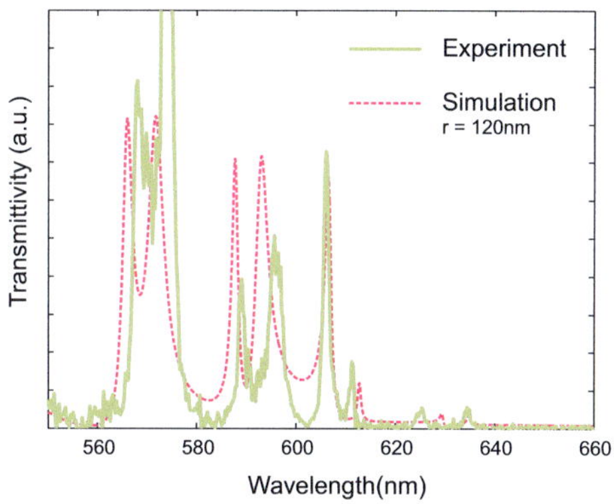

Figure 5.5: *Comparison of measured and calculated transmittivity at the polar angle of $\theta = 5°$. The best fit is obtained for a hole radius of 120 nm.*

proposed in this work. For such structures one of the most important issues in quality control is the homogeneity over a large area. In this section first a visual inspection method, which is based on using crossed polarization filters, is proposed. As a second method a spectrometer is utilized to reveal inhomogeneities, which were not observable with the visual method.

5.3.1 Visual inspection

A simple and straightforward method is the visual inspection. This method is also widely used in todays industry in diverse industrial fields. In PCS fabrication the visual inspection could be a rapid and reliable method to monitor the quality, as the optical properties of such structures are more relevant than only topological information.

For this purpose GMRs are the ideal candidates to reveal inhomogeneities, as these resonances are strongly related to the geometrical parameters of the PCS. Furthermore, these resonances can be obtained by simple transmission or reflection measurements using a white light source.
Using conventional transmission or reflection procedures the resulting image is a superposition of light interacting with the PCS[4] containing the intended information and the white light not interacting with the PCS, which can be interpreted as a disturbing background. As GMRs are usually narrow-band resonances relative to the white light spectra, the impression of the observer is dominated by the background. However, the absence of the background light results in a color impression to the observer, depending on the spectral position of the GMRs.

A method used in today's fabrication processes to perform visual inspection of PCS is dark field microscopy. In dark field microscopy the unscattered light, i. e., the light, which is transmitted through the system, is excluded from the image. This is realized by using a condenser with a higher numerical aperture than the objective and a dark field patch stop. These elements are designed in a way that the illuminating light is guided around the objective and only the scattered light is collected by the objective. Applying this method to the inspection of PCSs means that white light from the condenser is able to excite GMRs in the PCS, but cannot enter the numerical aperture of the objective. Due to the fixed size of the dark field patch stop and the numerical aperture of the condenser, this method does not allow any adjustments of the GMR wave vector.

As already introduced in section 4.2 using orthogonal oriented polarization filters, with their right alignment to the PCS, only the GMRs are revealed. This configuration can be used for the inspection of PCS and overcome the problem with the non adjustable GMR wave vector. In the following this inspection method is discussed with the help of imperfect PCSs fabricated with an electron beam lithography step. These PCS consist of a Nb_2O_5 layer on a quartz substrate with holes in a triangular geometry. The PCS fields are 150 μm x 150 μm squares, composed of 25 smaller squares with 30 μm x 30 μm.

[4] Light interacting with the PCS are GMRs.

These fields are written one by one, while only a pure deflection of the electron beam is used for each field. To write the next field a translation stage is used.

To capture the visual impression a single lens camera[5] was used, which was plugged to one of the exit ports of the microscope setup, depicted in Fig. 4.1. Since no pinhole is placed in the optical pathway of the used exit port, these pictures were recorded in a non-confocal configuration.

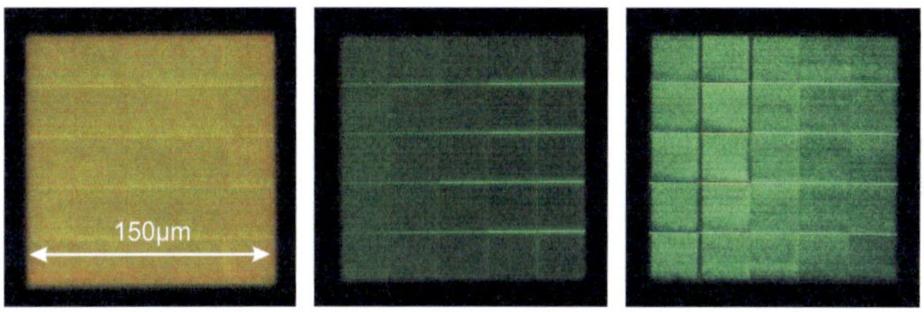

Figure 5.6: *Visual inspection of an imperfect PCSs using crossed polarization filters. The PCS fields are composed of 25 30 μm x 30 μm PCS fields. Stitching errors are clearly visible for each PCS field.*

In Fig. 5.6 three examples of PCSs captured with a magnification of 40x and an open slit-diaphragm, which means a continuum of wave vectors from 0° to 17°, are shown. Because of the transmission of the GMRs and the suppression of the background light, each PCS shows a specific color, depending on its geometrical parameters, such as hole radius or periodicity. On the other hand areas without a periodic structure appear dark. As discussed in section 2.6 one of the fabrication errors using electron beam lithography are stitching errors. Stitching errors occur when the distance of the PCS fields are either too big or too small. Hence at the edge of each PCS field the periodicity of the holes is different than in the rest of the PCS. This can lead to different phenomena at the edge of each PCS. With a different periodicity of the holes the spectral position of the GMRs will shift and hence the color for the observer will change. Another phenomenon, which is observed more often is scattering of modes in the slab structure on these edges, which renders the edges brighter than the rest of the crystal.

In Fig. 5.7 another series of PCSs fabricated in the same way shows another type of error, which is typical for electron beam lithography, namely unwanted variation in exposure

[5]D50 *Nikon*

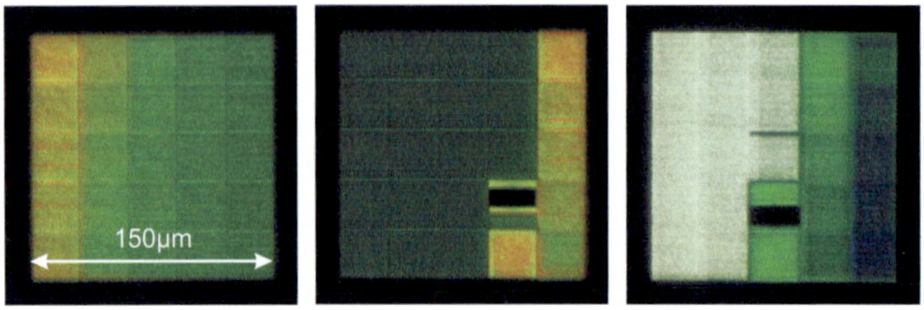

Figure 5.7: *Visual inspection of imperfect PCSs using crossed polarization filters. These examples show besides the stitching errors another common problem with electron beam lithography: namely exposure dose drift. This error results in a color drift of the PCS.*

dose. The hole size for the PCSs fabricated with electron beam lithography is regulated with the exposure dose. A higher exposure dose will change chemically more molecules in the photoresist, which yields a bigger hole size. Besides the stitching errors, in these three examples, a drift of the exposure dose is observable as a color drift.

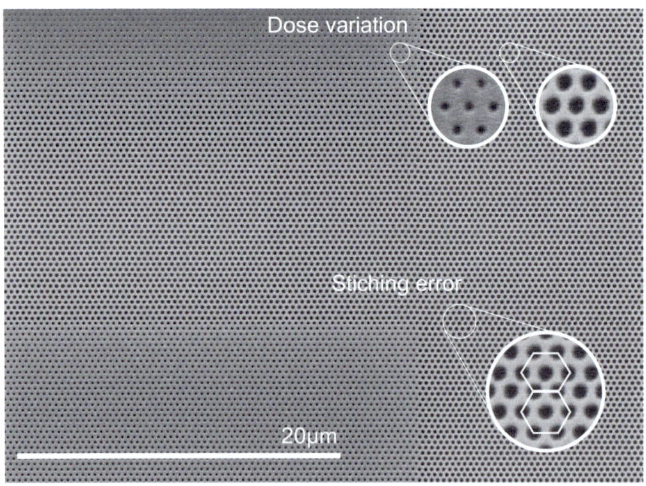

Figure 5.8: *SEM image of PCS shown in Fig. 5.7 left image. This image reveals the two discussed errors for electron beam lithography: mismatch of holes at the edge of each PCS field, which is the stitching error, and hole size drift, which originates from exposure dose drift.*

To verify these claims an SEM image of the PCS shown in Fig. 5.7 left image, is shown in Fig. 5.8. Here both errors are visible as topological errors: the stitching errors, as a mismatch on the edge of each PCS field and the exposure dose variation, as hole size variation. We notice that the periodicity of the structure can be aligned very well and remains stable during the writing process and do not contribute to the errors in the crystal structure.

5.3.2 Homogeneity investigations utilizing a spectrometer

In the previous section a visual inspection method was introduced, which allows a fast and low-cost quality control. The precision of this method is limited by the abilities of the human eye. In many applications, however, a more accurate and reproducible method is required. In this section a method is presented, which uses a spectrometer to analyze the GMRs. Performing spectrally and spatially resolved transmission measurements employing orthogonally oriented polarization filters, a direct access to GMRs is possible [112]. The advantages of this method are discussed using three examples. In the first two examples we characterize and compare the optical homogeneity of photonic crystal slabs fabricated by electron beam lithography and by laser interference lithography. Furthermore as a last example a complete scan across a PCS is shown. Here a GMR drift of 1.3% is revealed. To perform these experiments the microscope setup introduced in section 4.1.1 was used.

The PCS already shown in Fig. 5.7 left image, is analyzed in more detail utilizing a spectrometer. To measure a spatially limited spot, we use a magnification of 40x, which yields in combination with the pinhole (100 μm) a spot size of 2.5 μm on the sample. Furthermore, the slit-diaphragm was closed to record only GMRs at normal incidence. In Fig. 5.9 transmission spectra at three points[6] on the sample are shown. Due to the hole size differences, as discussed in the previous section, GMRs in point A and B show large differences. Since the hole size is bigger in point B compared to point A, the GMRs shift towards shorter wavelengths.

To demonstrate the limits of this method we compare now the transmission spectra in points B and C, which lie close to each other and produce almost the same color impression to an observer. These spectra are shown in Fig. 5.9 right graph. In this graph both spectra seem to overlap completely. To see their difference, however, we need to zoom in. The green shaded region in Fig. 5.9 is magnified and shown in Fig. 5.10. Here again the shift towards shorter wavelengths is observed, which is again interpreted as a

[6] Defined in the left picture in the same figure

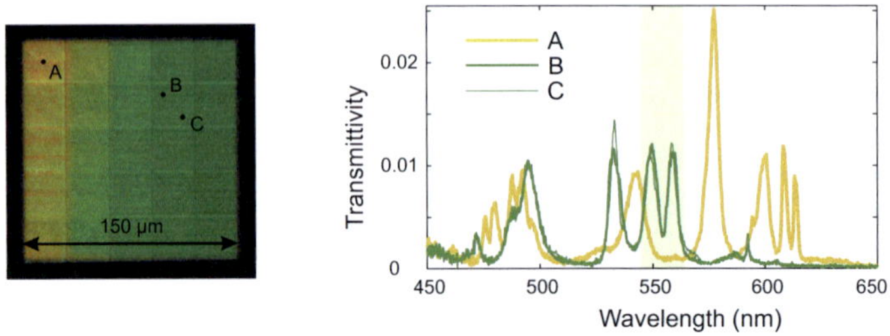

Figure 5.9: *Homogeneity inspection of a PCS fabricated with electron beam lithography utilizing a spectrometer. Left image: transmission image with crossed polarization filters. The drift of the hole size cause a color drift from yellow to green. Right graph: transmission spectra for different position on the PCS in a confocal configuration.*

drift of the hole size towards bigger hole sizes.

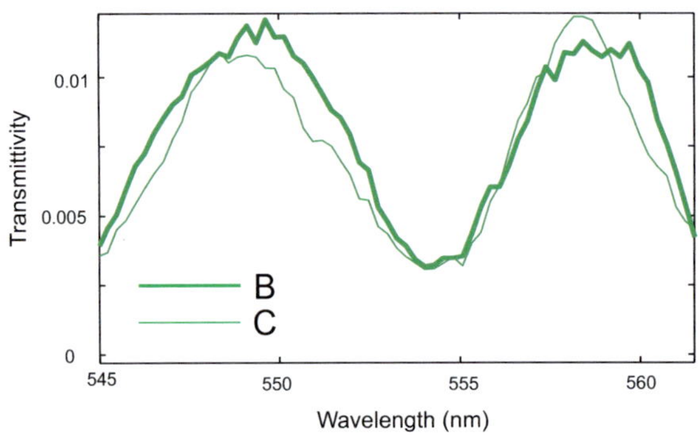

Figure 5.10: *Spectral zoom in the graph shown in Fig. 5.10. A spectral shift of less than 1 nm for the GMRs is observed.*

These investigations are continued on a PCS fabricated with laser interference lithography. Again a transmission image (Fig. 5.11 left image) is used to gain an overview of the homogeneity of the structure. Here an over modulation of the PCS structure, which shows itself as an intensity modulation, is clearly visible. This is caused by unwanted interferences and should not influence the periodicity[7] throughout the structure. This

[7]The periodicity is only a function of the wavelength of the used laser and the adjusted angle of the two beams.

assumption is approved by transmission measurements in point A and B in Fig. 5.11 (right graph). There is a difference in the intensity, but not in the spectral position, which indicates no periodicity modulation.

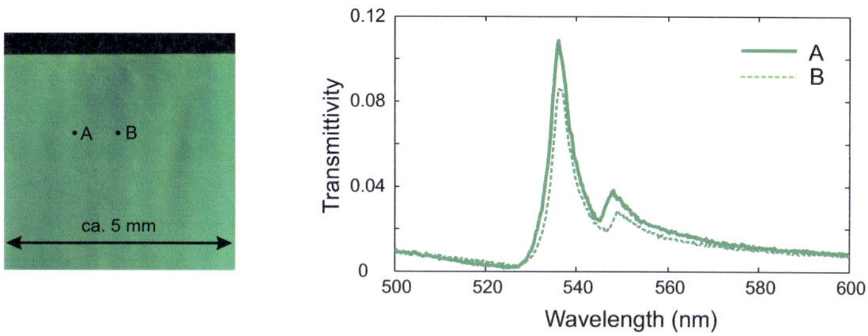

Figure 5.11: *Homogeneity inspection of a PCS fabricated with laser interference lithography utilizing a spectrometer. Left image: transmission image with crossed polarization filters. An intensity modulation is visible, which indicates unwanted interferences during the fabrication process. Right graph: transmission spectra for different positions on the PCS in a confocal configuration. The intensity modulation is verified and almost no spectral shift is observed.*

As a last example again a PCS fabricated with electron beam lithography is considered. The main focus of this investigation is the homogeneity across the sample [113, 114]. First the transmission image is recorded and shown in Fig. 5.12. To clarify the inhomogeneity of the PCS throughout the sample a false-color image of the red image channel of the image is prepared, too. Here the intensity of the red channel disappears at the upper end of the PCS field. This indicates a mode shift towards shorter wavelengths, which moves completely out of the red channel.

For the evidence of this assumption and to quantify the mode shift, a spectrally and spatially resolved transmission scan over the entire PCS field (150 μm) with 12.5 μm step size was recorded. The dotted line in Fig. 5.12 left image indicates the scan direction. For this measurement a magnification of 6x was chosen, which yields an optical resolution of ca. 16.6 μm. This scan reveals a drift of totally 7 nm of GMRs over the entire PCS.

This drift of 1.3 % illustrates in an impressive way the advantage of this method. Such a small paramter drift over a distance of 150 μm is difficult to observe with methods imaging only the topological properties, such as SEM.

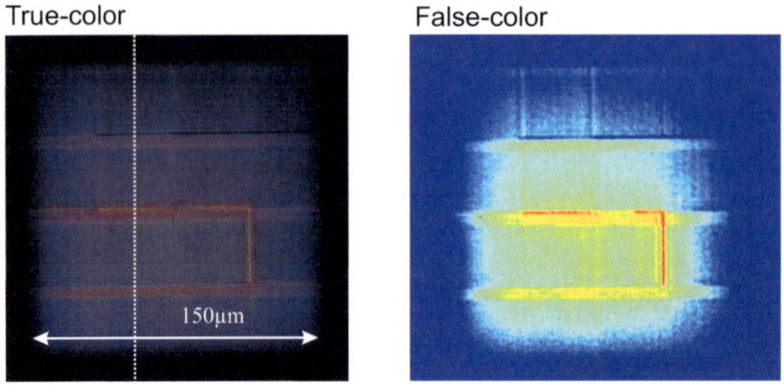

Figure 5.12: *Transmission image of a PCS fabricated with electron beam lithography using crossed polarization filters. Left image: a true-color image, which reveals stitching errors and indicates parameter drift. Right image: a false-color image of red image channel, indicating mode shift towards shorter wavelengths.*

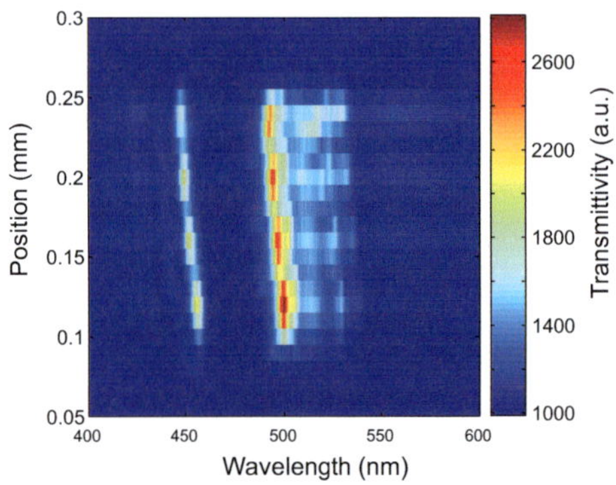

Figure 5.13: *Spectrally and spatially resolved transmission measurements at normal incidence. This scan reveals a 1.3 % drift in GMRs and illustrate the advantage of this method.*

5.4 Band diagram measurements above the light line

The band diagram of photonic crystals is the most powerful tool for scientists working with these structures. These graphs contain almost all optical information[8] relevant for designing optoelectronic devices using PCSs. For instance in applications such as distributed feedback (DFB) lasers [106, 115–117], light extraction from LEDs [74, 76, 77] or bandpass filters [118, 119], just to mention some of them, the knowledge about the

[8]This is true if the linewidth and the intensity of each mode is included in the band diagram, too.

band diagram is essential. In [120], Ganesh el al. use in an impressive way the band diagram of a PCS to design a surfaces to enhance the fluorescence emission of quantum dots. For both, the excitation of the quantum dots and the emission of light, GMRs of the same PCS are used to optimize these processes.

As we saw in section 2.3 the band diagram of a PCS can be divided into two regions. The region below the light line contains only guided modes, while the region above the light line provides the QGMs. The nature of guided modes complicates their measurement, as they propagate in the slab and are evanescent outside the slab. However, there are methods, which allow with additional effort to have access to these modes [62, 86]. In attenuated total reflection (ATR) spectroscopy, for instance, a high index prism is placed close to the PCS surface. This allows an efficient coupling of the evanescent waves of the totally reflected excitation light and guided modes in the PCS. On the other hand QGMs have an intrinsic loss to the far field and are accessible through their corresponding GMRs. Generally, transmission or reflection measurements are used to map the band diagram above the light line [103, 121–123]. The observed band diagrams are either an accumulation of dots, standing for the spectral position of the GMRs or an apposition of spectra, which contain not only information about the GMRs but also about the thin film interferences.

We propose to use crossed polarization filters for these measurements to obtain a band diagram containing information only about GMRs.

5.4.1 Measuring the band diagram in different crystal directions

The setup introduced in section 4.1 is used to perform all following band diagram measurements. Further, the area, which is recorded is limited due to the confocal setup. The combination of the 100 μm pinhole with the magnification of 40x, which was applied for all band diagram measurements in this section, limited this area to a spot of 2.5 μm. As the propagation lengths of QGMs are few μm, a higher spatial resolution will not offer any improvements. This even would reduce the recorded numbers of photons and worsen the signal to noise ratio (SNR). We exploit the whole angular range of the setup, which was 4° to 17°, to record 56 transmission spectra (exposure time 1 s), while opening the diaphragm using a step motor. As the calibration curve shown in Fig. 4.4 possesses a linear relation between the opening angle and the motor steps, we skip any post processing of the angular distribution. Applying the differential method introduced in section 4.1.2 we obtain 55 transmission spectra at defined angles. Each transmission spectrum has an angular distribution of approximately 0.23°, which is caused by the discrete number of angular steps. As a last step each spectrum is normalized to the spectrum of the light source, which is a halogen lamp, and is plotted as a function of the angle.

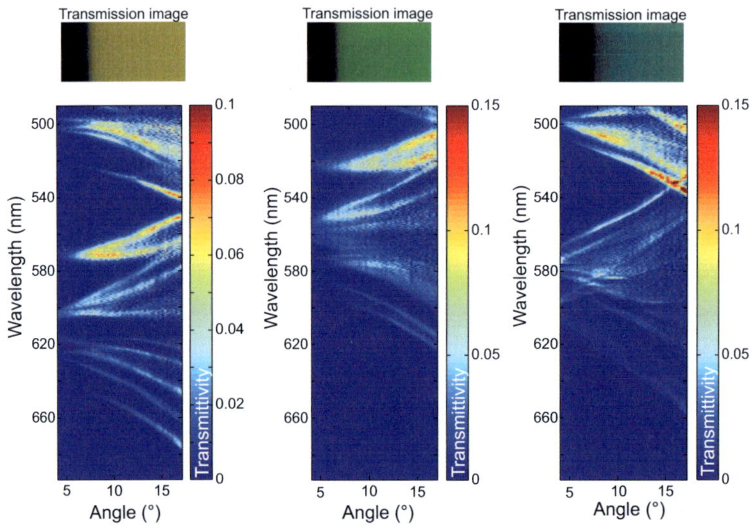

Figure 5.14: *Angle resolved transmission measurements applying a differential measurement method. Three different triangular PCSs with the periodicity of 360 nm, 330 nm and 290 nm are shown here. The corresponding transmission image using crossed polarization filters is shown above each graph. As expected a spectral shift of GMRs towards lower wavelengths with smaller periodicity is observed.*

In Fig. 5.14 measurements of three PCSs are shown. These photonic crystals consists of holes in a Nb_2O_5 layer of 221 nm [9] in a triangular geometry. They offer periodicities of 360 nm, 330 nm and 290 nm, respectively. As expected we observe a spectral shift of GMRs towards lower wavelengths with smaller periodicity.

For our further investigations on PCS's band diagrams and later in the next section the Q-factor of GMRs, we use the PCS introduced in section 5.2. As already geometrical and material properties of this PCS are well known, experimental results can be verified by theoretical calculations. In Fig. 5.15 a band diagram of this PCS is shown. A variety of modes with different spectral positions and linewidth are observed. To ensure that

[9] This value was obtained by deduction in section 5.2.1

the spectral resolution of the system is resolving all details of the GMRs, such as the linewidth of the modes, band diagram measurements were repeated with higher accuracy. These additional measurements were recorded with spectrometer gratings of 1200 lines per mm (l/mm) and 2400 l/mm, respectively.

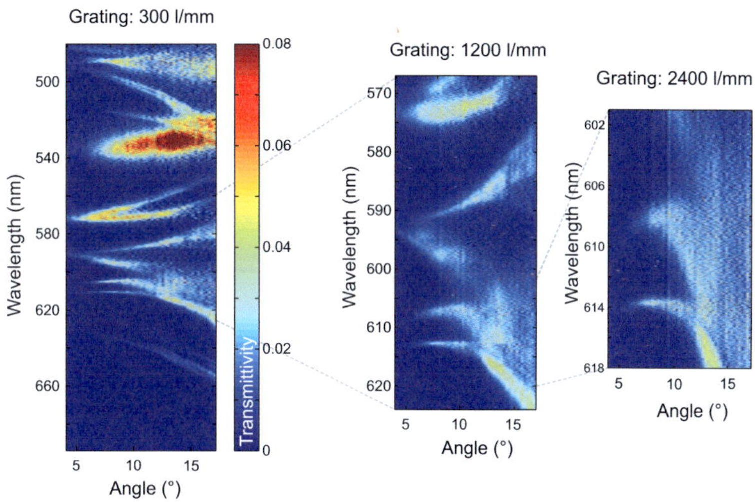

Figure 5.15: *Angle resolved transmission measurements applying a differential measurement method of the PCS introduced in section5.2. Furthermore transmission measurements with higher accuracy are performed to ensure the correct representation of the GMR with small linewidth. A blurring of the GMRs particular at high angles indicate a systematic error in the measurement technique.*

In these measurements we observe a blurring of the modes, especially for larger angles. This effect is not caused by the modes and their linewidth variation. The origin of this blurring is a systematic error in the measurement technique. As no slit-diaphragm was used as proposed in section 4.1.2, the observed band diagrams contain modes of all crystal directions. Since the PCS under investigation has a triangular geometry the slit-diaphragm proposed in Fig. 4.3 is applied. However, using the slit-diaphragm implicates higher SNR, due to the shadowing of the light source. To overcome this problem we implement a series of changes in the measuring procedure:

- The most straightforward change is the integration time itself, which was increased to 10 s. For higher integration times, however, the probability of detecting *cosmic rays*[10] grows too. Cosmic rays cause pixel error, which show up as a sharp peak in the spectra. This problem, however, is reduced by a filter in the analysis software.

- As a next change, the step size of the step motor for controlling the angle is increased. In a differential method this leads to an increase of collected photon numbers. For the following band diagram measurements 28 spectra were recorded. Here we have to accept a reduced angular resolution.

- In the previous measurements the high index layer with the nanostructure shows towards the condenser. If the sample is turned upside down the light form the condenser will pass the quartz substrate to reach the PCS field. Due to the refraction at the air quartz interface, the illumination angle will change to lower angles. Applying Snell's law, we obtain an angular illumination from $3°$ to $13°$. With 28 spectra distributed over this angle range, we obtain an angular resolution of $0.37°$. Turning the sample upside down did not result in a higher number of collected photons, but in a higher angular resolution.

- The width is of the slit-diaphragm allows to collects more photons, too. A too broad slit width, however, transmits also other illumination directions than desired. To have a balanced trade-off between collected photon number and illumination direction for the following measurements a slit width of 2 mm was chosen.

The impact of using the slit-diaphragm is apparent in Fig. 5.16. Here the PCS shown in Fig. 5.15 was measured again with the discussed changes. Clearly the two crystal directions ($\Gamma - K$ and $\Gamma - M$) can be separated and no mode blurring is observed any more. This kind of measurement reveals the variety of GMR's angular dependence and linewidth.

To be able to compare these band diagrams with band diagrams calculated as a function of the wave vector, as exemplary shown in section 2.3.3, the angle θ can be derived from the relation of the in plane wave vector ($k_{inplane}$) with the complete wave vector (k_0) as the following shows:

$$\sin\theta = k_{inplane}/k_0. \tag{5.1}$$

With $k_0 = \frac{2\pi}{\lambda}$ the angle θ is a function of $k_{inplane}$ and the wavelength of the mode.

[10]Cosmic rays are energetic particle originating from outer space.

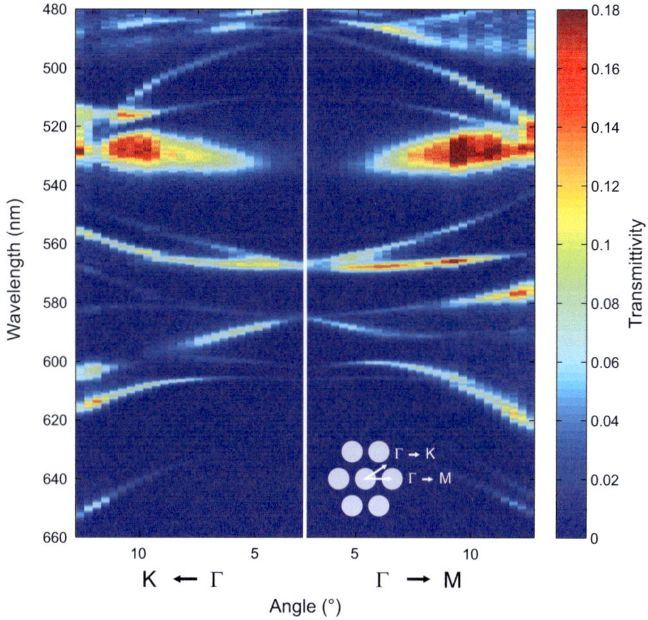

Figure 5.16: *Band diagram measurement above the light line in* $\Gamma - M$ *and* $\Gamma - K$ *direction. Here a slit diaphragm is used to separate the two directions in transmission measurements. This band diagram do not only reveal the angular dependence of GMR's spectral position, but also their linewidth.*

5.4.2 Verifications utilizing FDTD simulations

In spite of the caution to have a measurement technique with no systematic errors, the results of the previous section need to be verified by simulations. The angular broadening of each angle step of 0.37° may cause for highly dispersive modes a broadening of the mode linewidth. For this purpose the FDTD method introduced in chapter 3 in particular section 3.3 is applied. The geometrical and material parameters of the PCS under investigation known from section 5.2 are implemented into the simulation cell. Here a plane wave source with a polarization direction of 45° in respect to $\Gamma - M$ direction was used, which could excite both the s- and p-polarization in one simulation run. Both polarization directions are needed for this comparison as the experimental configuration delivers both, too (see section 4.3).

The simulation results are shown in Fig. 5.17, which is the transmittivity of the s- and p-polarization in $\Gamma - M$ direction from 550 nm to 660 nm.

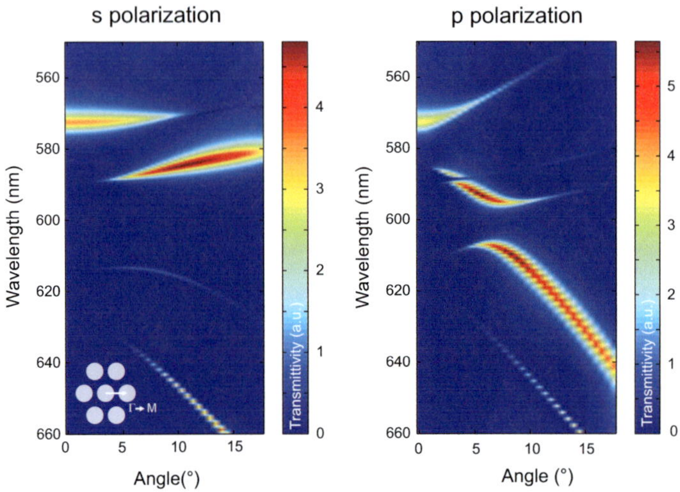

Figure 5.17: *Transmittivity calculations utilizing a 3D FDTD simulation method in s- and p-polarization. Since the illumination source is separated from the GMRs in the time domain and filtered out, only GMRs are transmitted.*

This limited spectral range is due to a general problem with FDTD calculations for this kind of simulations for PCS transmittivity. At some specific frequencies artificial resonances are superimposed with the GMRs. These resonances have a much higher intensity than GMRs and render the simulations impossilbe. Their origin lies in the imperfect absorption of PMLs at high angles of incidence. However, in a transmission process through a PCS part of the light might be diffracted and hit the PML at the bottom or top of the simulation cell under a high angle. As the PML cannot absorb it and the walls of the simulation cell are reflecting, too, this light will generate a resonance, which is detected by the monitor point.
A solution to this problem is to increase the number of the PMLs. This could decrease the intensity of these artificial resonances, but did not helped to eliminate them completely. Hence for this problem no solution could be found in the scope of this thesis and limitations were accepted.

Combining the results from Fig. 5.17 to one graph and plot from 3° to 13° we obtain the left plot in Fig. 5.18. We plot the experimental result for the same spectral range and crystal direction in Fig. 5.18 right plot. Good agreement for both, resonance angular dependence and linewidth, is observed.

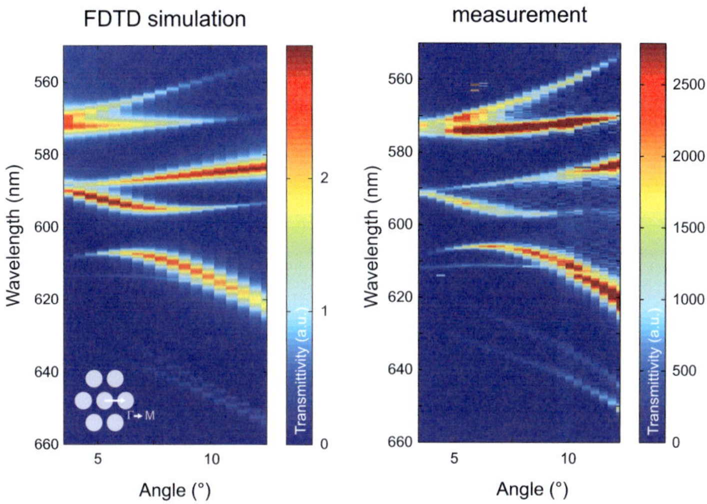

Figure 5.18: *Verification of measured band diagram above the light line in* $\Gamma - M$ *direction with 3D FDTD simulations. A good agreement for both, resonance angular dependence and linewidth, is observed.*

5.5 The angle and spatially resolved quality factor

Besides the spectral position of modes their Q-factor, which is inversely proportional to the linewidth, is one of the most interesting and important parameter of photonic crystal modes. As we learned from the previous section the proposed characterization method, which is using crossed polarization filters, does not only reveal the spectral position, but also the linewidth of the GMRs. GMRs show a Lorentzian line shape.

The Q-factor of GMRs is a function of their wave vector and a function of position for an imperfect PCS. In this section both issues are discussed.

5.5.1 Experimental quality factor determination

The Q-factor is a dimensionless parameter of an oscillating system, which specifies the damping of the system. A high Q-factor corresponds to a lower rate for the energy loss of the system, which results in a longer duration of the oscillation. In terms of GMRs, which exhibit an intrinsic loss mechanism, the Q-factor is a measure of its losses or equivalently the lifetime of a resonant photon in the PCS. Generally the Q-factor is defined as:

$$Q \propto \frac{\text{Energy stored}}{\text{Energy dissipated per cycle}}. \tag{5.2}$$

Besides this definition the Q-factor is equivalently defined as the ratio of the resonant frequency and the bandwidth of the resonance. Here we calculate the Q-factor as:

$$Q = \frac{\lambda_0}{\text{FWHM}}, \tag{5.3}$$

where λ_0 is central wavelength and FWHM[11] is the linewidth of the resonance.

To obtain the linewidth of the GMR under investigation, the difference between the measured spectra and a theoretical description of the resonance is minimized. Since the GMR possess a Lorentzian line shape the following term is applied to express the intensity transmission spectrum:

$$t = f_0 \frac{\gamma}{i(\lambda - \lambda_0) + \gamma}, \tag{5.4}$$

where λ_0 is the central wavelength and γ the width of the GMR. f_0 is the amplitude of the GMR. In Fig. 5.19 exemplary results of the fitting procedure is shown. The same mode is measured at 3° and 10° using the differential method with crossed polarization filter. A good overlap of the experimental results with the fitted curve is observed.

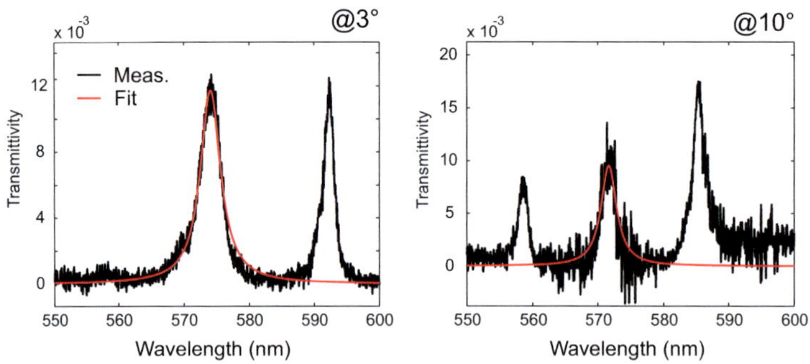

Figure 5.19: *Two examplary transmission spectra at 3° and 10° and their corresponding fitted curves. A good overlap of the measured data and a Lorentzian line shape is observed.*

Since γ is 1/2 FWHM we end up with the following term for the Q-factor

[11] Full width at half maximum

$$Q = \frac{\lambda_0}{2\gamma}. \quad (5.5)$$

Comparing equation 5.4 with equation 2.10 here a more simple and intuitive term is used for the fitting procedure. Moreover less paramters are used for the fitting. This allows for an easier to implement fitting procedure for the Q-factor determination.

In the scope of this work an analysis software was developed in the *Matlab*[12] environment. The requirement for this program was the import and plotting of angle resolved transmission data, extraction of individual modes, and their automatic Q-factor determination. As for the automatic Q-factor determination the intervention of the user is unavoidable a graphical user interface (GUI) was developed.

5.5.2 Angle resolved quality factor determination

The PCS with the band diagram shown in Fig. 5.16 is again the object of our investigations. To ensure better conditions for the Q-factor determination process, the angle resolved transmission measurements were performed with higher accuracy, here the spectrometer grating with 1200 l/mm was used.
Three angle resolved measurements in the $\Gamma - M$ direction for different spectral ranges were performed (Fig. 5.20, 5.21, 5.22). These figures show the large variation in Q-factors for GMRs as a function of the angle.

In Fig. 5.20 the left image depicts the transmission spectrum with four modes. The angular dependence of the Q-factor is shown in the same figure on the right-hand side. The Q-factor for some modes cannot be determined for all angles, as they either disappear or merge with other more dominant modes. This happens for example for mode 3, which is less intense compared to mode 4 and could not be separated from it.

Investigating mode 4 in more detail we observe an angular evolution of the Q-factor, which shows a maximum at about 5° with a value of 500. However, for higher angles the Q-factor drops down to 100. To the best of our knowledge this behavior was never observed in a PCS before. Recently a DFB laser was shown [124] with laser action at 3°, this was explained by the Q-factor of the GMR contributing to the laser action. It was shown theoretically that this mode has a maximum in Q-factor at around 3°. Our experimental results deliver the experimental evidence for this behavior.

[12] Matlab is a commercial numerical computing environment developed by *The MathWorks*.

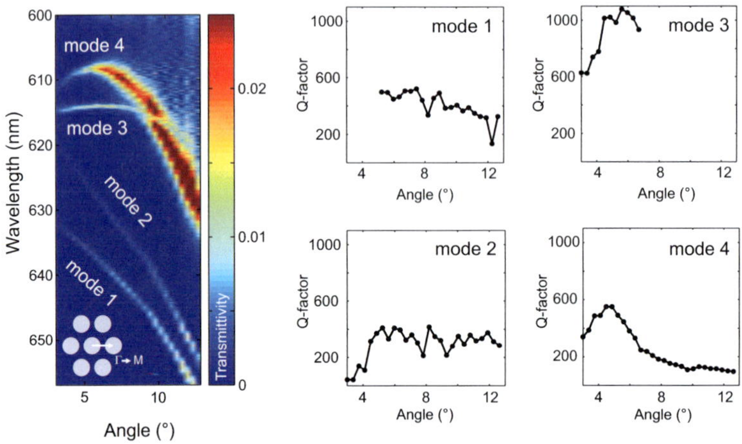

Figure 5.20: *Angle resolved Q-factor analysis in $\Gamma - M$ direction of the PCS with the band diagram shown in 5.16 with the central wavelength at 630 nm. Left image: the band diagram with 4 modes. Right graphs: Q-factor angular dependence.*

These measurements reveal another interesting behavior of GMRs. We observe for modes with a higher Q-factor a lower intensity and vise versa. Comparing for instance modes 3 and 4 or modes 9 and 10 this effect is clearly visible. Even in the angular dependence of a mode, when its Q-factor changes the intensity follows this rule. This is notably visible for all modes in Fig. 5.21. Although the intensity and the linewidth of a GMR show not a simple relation [125] this observation can be used as a rule of thumb. Photonic crystal modes with a high Q-factor, which have a small linewidth, will have low coupling to the far field modes. Thus, in transmission measurements excitation of these modes and out-coupling again to far field modes are inefficient and these modes appear darker. On the other hand modes with a low Q-factor show a strong coupling to far field modes due to their higher intrinsic losses. Hence, they have a larger contribution to their GMR's intensity.

5.5.3 Spatial resolved quality factor determination

In section 5.3.1 we investigated the inhomogeneity of PCSs by analyzing their GMR's spectral position. The GMR's spectral position is a function of its geometrical parameters and its drift is a good indicator for geometrical parameter drift. However, there might be

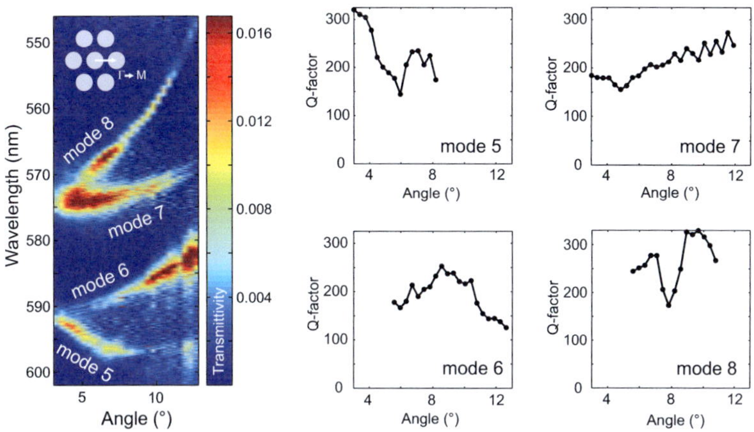

Figure 5.21: *Angle resolved Q-factor analysis in* $\Gamma - M$ *direction of the PCS with the band diagram shown in 5.16 with the central wavelength at 575 nm. Left image: the band diagram with 4 modes. Right graphs: Q-factor angular dependence.*

parameters of the PCS, which affect the Q-factor but not the spectral position of GMRs. These parameters might be surface roughnesses or material impurities. Since for some applications, such as photonic crystal lasers [124] the Q-factor plays a significant role, homogeneity analysis should be performed utilizing the Q-factor itself. Hence in this section we perform spatially and angle resolved Q-factor measurements.

We perform angle resolved transmission measurements in $\Gamma - M$ direction of the PCS under investigation at 7 positions. The measurement points have a distance of 4 μm [13] and were starting on one of the edges of the PCS field. Because of the fabrication procedure, generally the edge of PCS fields show slightly different geometrical properties. Fig. 5.23 depicts the angular dependence of mode 3 and 4 for these measurement points.[14] These pictures show a small spectral position drift of the modes, but no clearly visible linewidth variation.

For the further analysis the Q-factor of mode 4 for each position is determined and shown in Fig. 5.24. For position 7, which is more inside the PCS field we observe a angular dependence of the Q-factor comparable with the result in Fig. 5.20. Here we have a

[13]Again the 40x magnification was used, which results in a spatial resolution of 2.5 μm.
[14]For these measurements the spectrometer grating was 2400 l/mm.

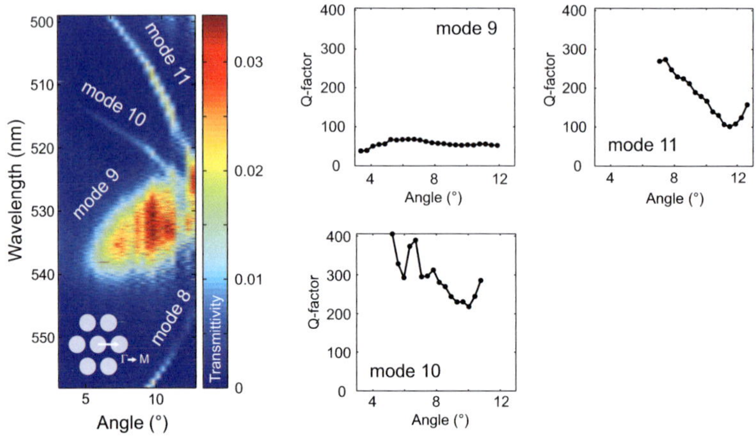

Figure 5.22: *Angle resolved Q-factor analysis in $\Gamma - M$ direction of the PCS with the band diagram shown in 5.16 with the central wavelength at 530 nm. Left image: the band diagram with 3 modes. Right graphs: Q-factor angular dependence.*

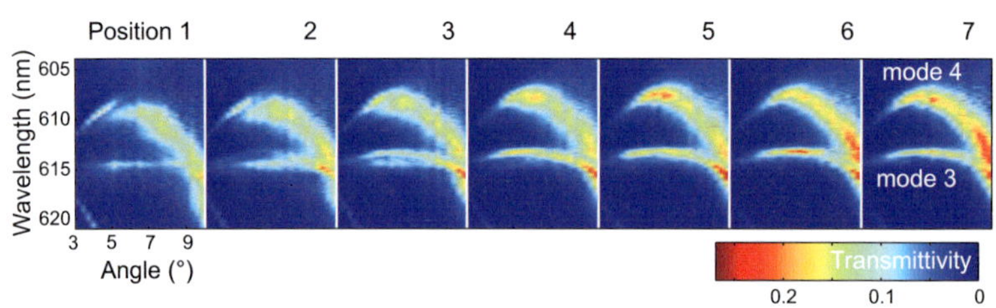

Figure 5.23: *Band diagram of mode 3 and 4 at different positions starting on the edge of the PCS. The distance of these positions is 4 μm and a spectrometer grating with 2400 l/mm was used to obtain high accuracy transmission spectra. A slightly drift of the GMR's spectral position is visible.*

maximum of the Q-factor at 5°. Approaching the PCS field edge the maximum of the Q-factor, however, shifts to smaller angles.

To visualize this Q-factor drift and map the homogeneity of the PCS we performed a transmission scan at 6°. At this particular angle high variation in Q-factor is expected from Fig. 5.24. For this scan we used an additional 1.5x magnification lens in the

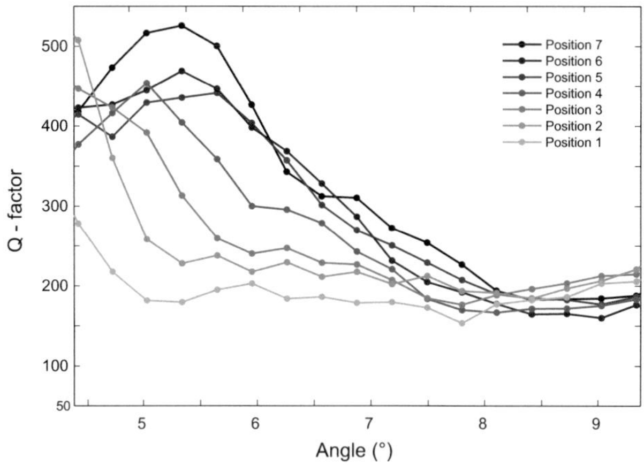

Figure 5.24: *Q-factor angular dependence of mode 4 for band diagrams shown in Fig. 5.23. The Q-factor curve shows a maximum at about 5° for a position deep in the PCS field. This maximum drift towards lower angles for position closer to the PCS edge.*

detection arm to increase the spatial resolution to approximately 1.6 μm. Again to obtain the transmission at a particular angle we applied the proposed differential method and performed two scans of the same PCS area with two different opening angles for the slit-diaphragm. In a post processing step we calculated the difference of these spectra in each pixel and obtained the GMRs for each position. A fitting routine determined the spectral position and the Q-factor of mode 4.

These results are shown in Fig. 5.25 and 5.26. While the resonance wavelength varies from 607 nm to 609 nm ($< 0.5\%$), we observe for the Q-factor a lowering from 350 down to 100 ($\approx 70\%$). With this method small geometric variations, which can affect the fundamental functionality of optoelectronic devices, can be visualized.

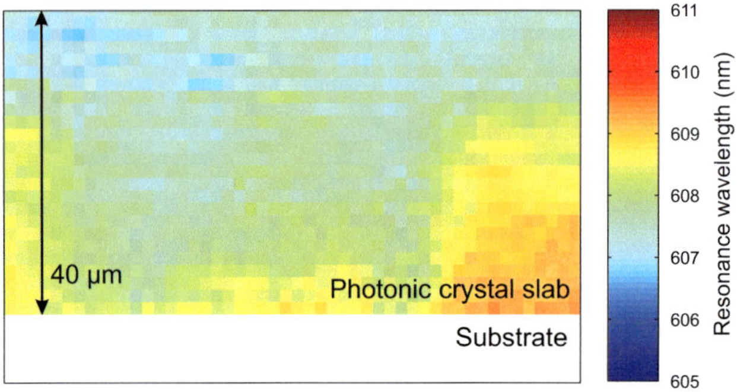

Figure 5.25: *Resonance wavelength gained by transmittivity scan at 6° around the edge of the PCS field. For this scan an additional magnification lens was used, which increased the resolution to about 1.6 µm. Here a variation in the resonance wavelength from 607 nm to 609 nm is observed.*

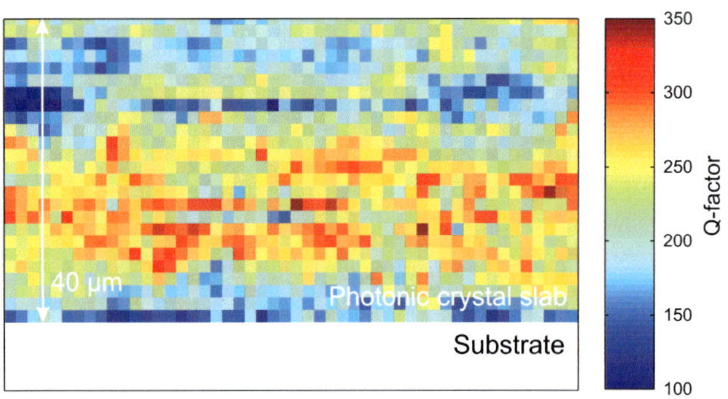

Figure 5.26: *Q-factor of mode 4 gained by the same transmittivity scan shown in Fig. 5.25. Although no significant changes in resonance wavelength are observed, the Q-factor shows large variations from 350 down to 100.*

Chapter 6

An optoelectronic biosensor using photonic crystal slabs

Summary

PCSs are known to be utilized as transducers in the biotechnology, in particular as biosensors. In this chapter we propose a method of measurement in biosensing based on the direct access to GMRs with the potential to be cost-efficient and compact. This method requires solely an LED and a photodiode, as light source and detector, respectively. The real-time label-free detection of the binding kinetics of a 2.5 nM solution of the protein streptavidin on a biotin-functionalized photonic crystal surface is demonstrated.

6.1 Motivation

In section 1.1 we learned about the most recent optical label-free methods of detecting biological and chemical substances. Although these methods are well established today, the development of cost-efficient and compact sensor systems, achieved with minimal apparatus complexity and resource expenses, still remains a big challenge. For instance devices based on SPRs detect a given analyte by spectrally analyzing the respective resonance. This procedure often requires sophisticated periphery, such as e.g. spectrometers and computational resources.

In particular one important application of such instrumentation is to investigate molecular interactions, such as protein-protein coupling or enzyme-based reactions, to assess binding affinities, kinetic processes, and reaction efficiencies, which are critical parameters for, e.g., biological research and drug discovery. Today's market offers various devices with these functionalities. However, they are a burden on the space and finances of research labs.

A cost-efficient solution would increase the availability of these devices and accelerate research in this field. On the other hand, flexible point-of-care applications would benefit from a compact device with a high sensitivity.

6.2 Resonance shift determination via intensity measurements

The resonance frequency of QGMs sensitively depends on the geometrical and material properties of the PCS. This fact was used for homogeneity characterization in section 5.3.1, where hole radius or shape was inhomogeneous across the PCS. However, the resonance frequency of QGMs responds very sensitively to changes in the refractive index of contributing materials or the presence of additional layers close to the PCS. A change in refractive index of a material penetrated by the QGM, results in a change of the effective refractive index for the QGM and thus a change in its resonance frequency.

It is straightforward to employ this relation for sensing applications based on PCSs. In particular in biosensors, the surface of the PCS is functionalized in a way that it can bind specific biological or chemical analyte substances. The presence of these substances might cause the desired refractive index change, if these substances have a different refractive index in their pure form than, e.g., water.

As the QGMs are directly related to their corresponding GMRs, a simple transmission or reflection measurement will reveal their spectral position. However, for this purpose a spectrometer and computational resources are still necessary. To solve the challenge of realizing a cost-efficient and compact detection method the problem can be reduced to the detection of the GMR shift.

In this chapter a technology platform for converting the GMR shift into an intensity is proposed. This method is ideal e. g. for detecting biological materials.

6.2.1 Convolution of a spectrally limited light source and guided-mode resonances

We propose a measurement scheme based on transmission through a PCS, employing crossed polarization filters. This allows for a pure transmission of the GMRs under investigation. Moreover, we use a light source with a spectrally limited spectrum possessing rising and falling edges. The convolution of GMRs with this light source results in a dependency of the intensity of the transmission on the spectral shift of GMR. This function shows a positive slope, when the GMRs overlap with the rising edge of the light source and a negative slope for GMRs overlapping with the falling edge of the light source. Therefore, a simple intensity measurement can replace spectral analysis of GMRs.

These considerations are summarized in an exemplary scheme depicted in Fig. 6.1. Here the spectrally limited light source is chosen to be an LED, where the GMR overlaps with its falling edge. An increase in refractive index at the surface of the PCS results in a shift of the GMR. The intensity of the light source is reduced at the new spectral position of the GMR and thus the transmitted intensity is decreased.

A simple setup as illustrated in Fig. 6.2 contains all required elements to realize the desired device. Here, a photodiode is utilized to detect the intensity of the transmission.

6.2.2 Bulk refractive index measurements

The most important optical element in this device is the PCS acting as the transducer. In this section we study the response of GMRs provided by the PCS to a bulk refractive index change. A bulk refractive index study delivers an estimation of the sensitivity of the

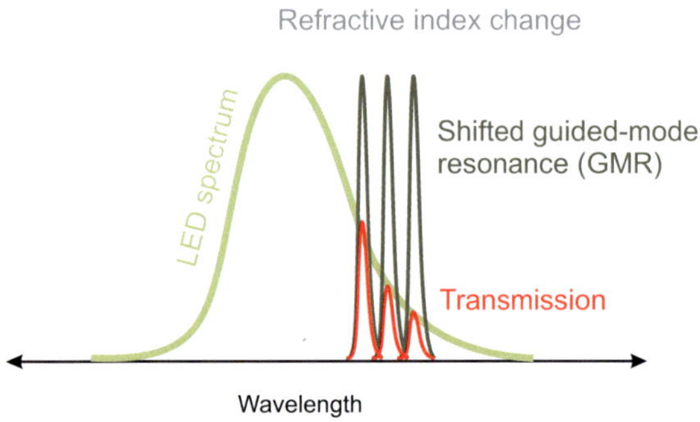

Figure 6.1: *Basic concept of the conversion of GMR shift into intensity change using an LED as the light source. The convolution of the LED spectrum with the GMR results in a function of the spectral shift of GMR versus the intensity of the transmission. Hence a simple intensity measurement suffices to determine the GMR shift.*

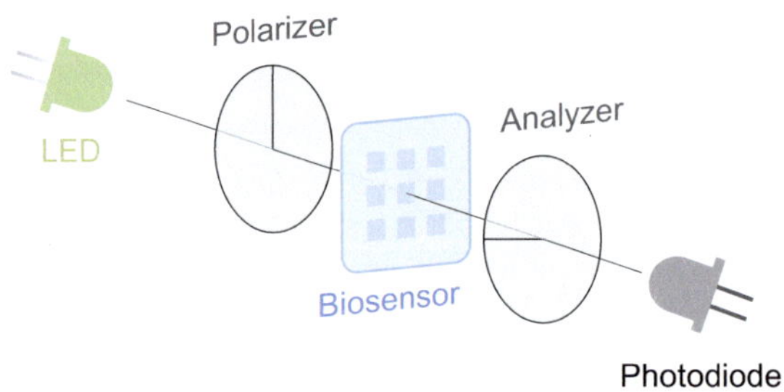

Figure 6.2: *Schematic concept of a compact and low-cost biosensor for affinity measurements in real time, based on detection of transmitted intensity changes due to GMR shifts. Here an LED is applied as the light source, while a photodiode is used for the intensity detection.*

system, and hence is a compulsory approach for selecting the proper PCS parameters. In a bulk refractive index measurement the refractive index of the liquid surrounding the PCS is tuned, while the GMR's shift is recorded and related to the refractive index of the liquid.

The PCS, chosen for this experiment, exhibits a linear geometry and was composed of a tantalum pentoxide (Ta_2O_5) layer on a glass substrate.[1] The periodicity of the PCS was $\Lambda = 370$ nm.

[1]The PCS was fabricated using a laser interference lithography step.

A common way to prepare liquids with tuned refractive index is the mixing of water and glycerin, both are biologically compatible liquids. However, inspired by the industrial sugar concentration determination, we chose a water-sucrose[2] solution. In the microscope setup as described in section 4.1.1 the transmission of this PCS is recorded using three different sucrose solutions: 0 g/ml, 0.5 g/ml and 1 g/ml. These concentrations correspond to refractive indies of 1.33, 1.435 and 1.54, respectively. For these measurements the diaphragm of the condenser was closed to obtain illumination at normal incidence. The light source was a halogen lamp possessing a wide spectral characteristic.

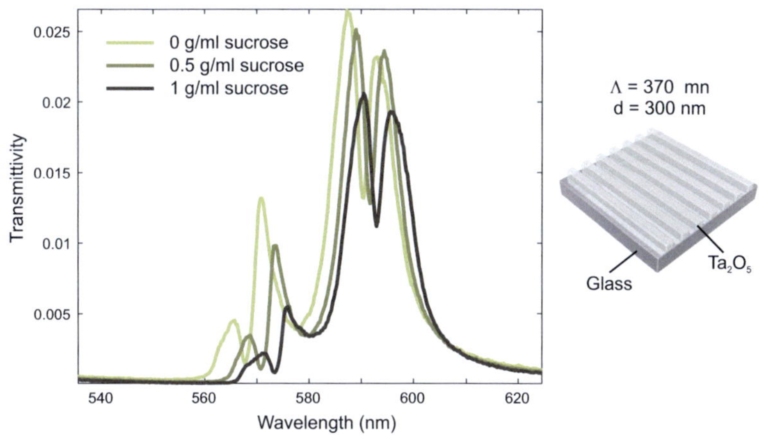

Figure 6.3: *Sensitivity determination of the PCS of interest using a bulk refractive index measurement. The light source is the built-in light source of the microscope (100 W halogen lamp), whose spectrum is used to normalize the transmission spectra. We determine an average sensitivity for all four GMRs of ca.* $\Delta\lambda/\Delta n$ *= 25 nm/RIU.*

In Fig. 6.3 the transmission measurements are shown. All four resonances shift similarly and contribute to the sensitivity of the sensor. We determined the sensitivity for the four measured resonances to be $\Delta\lambda/\Delta$n = ca. 25 nm/RIU (nm per refractive index unit) on average. In comparison with other optical methods, such as SPR based sensors, or other PCS this value is not satisfying [126]. In [127] Shi et al. have shown a PCS with a sensitivity of 327 nm/RIU for an individual mode operating at 600 nm. In the scope of this work, however, there was no access to PCSs with higher sensitivities.

[2]Sucrose is well known as table sugar with the molecular formula: $C_{12}H_{22}O_{11}$.

As a next step we perform transmission measurements as suggested, with an LED as the light source. This LED shows its maximum emission intensity at 518 nm. The spectral characteristics of the LED and the spectral position of GMRs was chosen in a way that GMRs overlap with the falling edge of the LED spectra. It is of high value that GMRs remain on this falling edge despite any resonance shift. To yield more information about the intensity modulation for each GMR, we first analyze the spectrum with a spectrometer, instead of a photodiode.

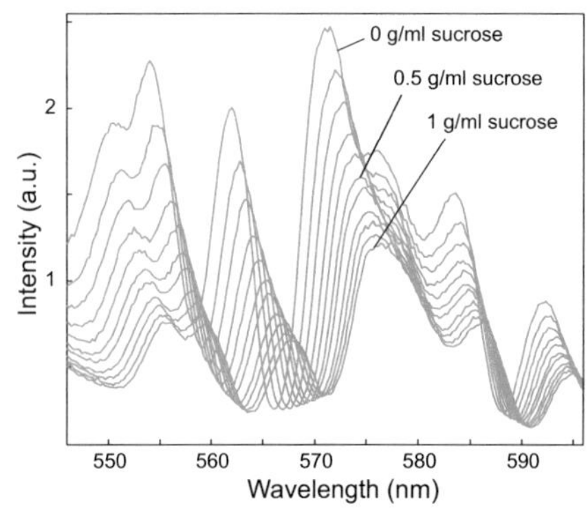

Figure 6.4: *Transmission measurements with tuned refractive index and an LED as the light source. An intensity drop of up to 75 % is observed, which stands in relation to the GMR shift.*

Again a series of sucrose solutions is prepared, which is used to cover the PCS surface. In Fig. 6.4 eleven measurements with sucrose concentrations from 0 g/ml to 1 g/ml with 0.1 g/ml steps are shown. Comparing these GMRs with the results obtained in Fig. 6.3, we observe differences in mode intensity and spectral position. These are a result of the limited LED spectrum and its radiation characteristics.
Because of the tuned refractive index GMRs show on average a shift of about 10 nm. Moreover, we observe as desired an intensity drop for all modes, which is up to 75% for individual modes.

The integration of the spectra over all GMRs yields intensity values comparable with measurements with the photodiode. In Fig. 6.5 this intensity curve is plotted over the associated refractive index. As expected the intensity of transmitted GMRs follows the falling edge of the LED spectrum and drops down to less than 45%.

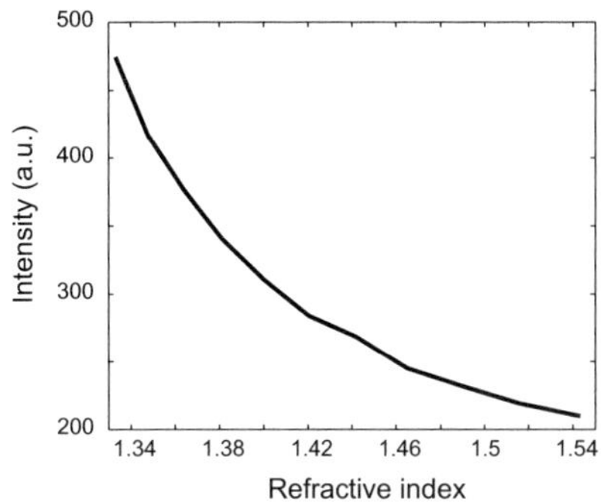

Figure 6.5: *Intensity over refractive index of the surrounding liquid obtained by integrating spectra obtained in Fig. 6.4. The decreasing intensity is due to the GMR shift, which is a function of the refractive index. As expected the intensity drop follows the falling edge of the LED spectrum.*

6.3 Design of a demonstrator

In the previous section the basic concept of the proposed method was proven in a bulk refractive index measurement. To demonstrate the ability of this method to be a compact and cost-efficient method, in this section a demonstrator of such a sensor is designed and realized.

6.3.1 Mechanical and electrical elements of the demonstrator

The setup of the demonstrator is as simple as depicted in Fig. 6.2.

The PCS, which is used in these experiments as transducer provides various GMRs between 540 nm and 600 nm, while surrounded by water or sucrose solution (Fig. 6.4). To guarantee that these GMRs lie in the falling edge of the LED the emission maximum of the LED was at 518 nm. To detect the transmitted intensity, which is a function of the GMR shift, we use a silicon photodiode. The LED as well as the photodiode offer an integrated epoxy optic. On the LED side this optic partly parallelizes the emitted light and directs it towards the detector. On the detector side the epoxy optic focuses the transmitted light into the detecting area.

In the following we discuss the driver electronics for the LED and the photodiode, the flow cell, which brings the analyte to the surface of the PCS, and the frame, which brings all components together.

The driver electronics

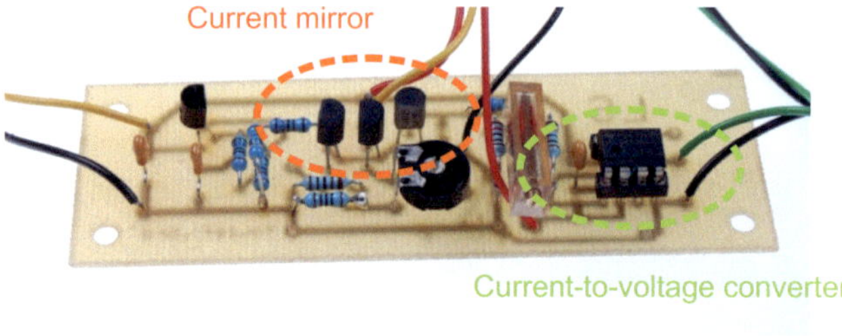

Figure 6.6: *Photograph of the driving electronics for the demonstrator integrated on one board. The LED current is delivered by a current mirror, while the photo current is converted into a voltage using a current-to-voltage converter. The output voltage is analyzed and recorded using a data acquisition device.*

The driver electronics can be divided into two parts: the LED control and the photodiode readout electronics. Both circuits are integrated on one board (Fig. 6.6) and use the same battery as power supply.

As the resonance shift determination is converted in an intensity measurement a stable luminous flux is an essential issue. For this purpose a current mirror is used, which is independent of the LED voltage. The luminous flux and the spectral characteristics of an LED are also a function of its temperature. In the scope of this work, however, we neglect variations caused by ambient temperature drift, as measurements were carried out in air-conditioned laboratories.

The photodiode is driven in a short circuit configuration with a current-to-voltage converter, since no fast measurements are intended and a low dark current is preferred. The output voltage is analyzed and recorded using a data acquisition device,[3] which was connected to a computer.

The flow cell

In the next section we will present a measurement of the coupling process of a protein to the functionalized surface of a PCS. To be able to generate reproducible results a flow cell plugged to the PCS surface is designed. As the coupling of the

[3]NI USB 6009, *National Instruments*

protein to be detected is very strong to the functionalized surface, each protein detection process is accompanied by a prior cleaning and functionalization process. Hence a fast exchange of the PCS is required. Furthermore, a cost-efficient solution was preferred.

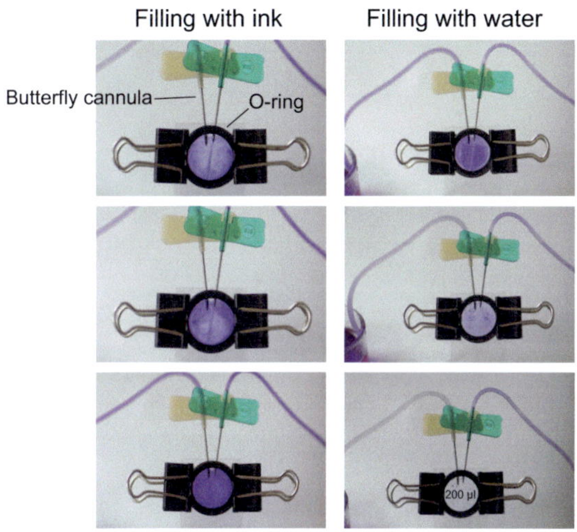

Figure 6.7: *Photograph of the flow cell during volume change with an ink-water solution and pure water. The flow cell consists of a o-ring squeezed between two glass substrates (one with the PCS on its surface). The liquid supply is realized using a butterfly cannula, which is pierced into the o-ring. The capacity of the flow cell depends on the diameter and the thickness of the o-ring and was in this case approximately 200 μl*

The flow cell consists of an elastic o-ring squeezed between a glass substrate and the PCS, which is fabricated on a glass substrate, too, using fold back clips. A liquid supply containing the analyte is realized using a butterfly cannula, which is pierced into the o-ring. Similarly for the outflow also a butterfly cannula is used. The pumping of the liquid is performed using a syringe through the supply cannula, while the outflow cannula is opened. The liquid capacity of the flow cell depends on the diameter and the thickness of the o-ring. In our case we have approximately a liquid volume of 200 μl.

For bulk refractive index measurements, for instance, the complete liquid exchange of the flow cell is of high impact. In Fig. 6.7 the flow cell is flooded in succession with a ink solution and pure water, to observe the exchange of the fluid. Only 3 ml fluid suffices for this purpose.

The frame

To bring all components of the demonstrator together in a compact way and guarantee correct adjustment, a frame was designed by CAD software[4] and fabricated using a 3D plotter.[5] The 3D plotter uses a polymer material for the fabrication and provides a resolution of about 0.1 mm.

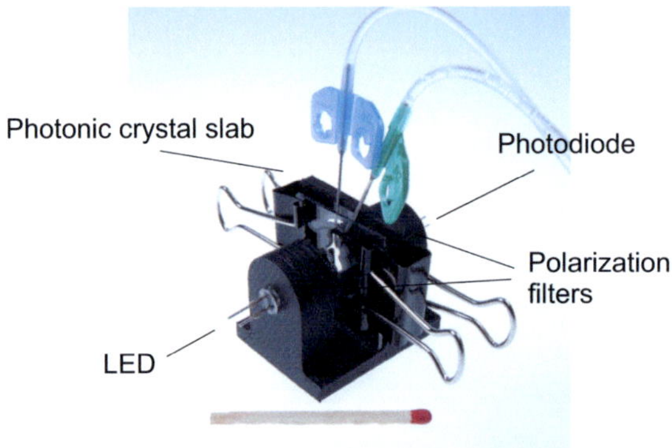

Figure 6.8: *Photograph of the assembled biosensor demonstrator.*

Fig. 6.8 shows the finished demonstrator with a match as a scale bar. Except of the PCS, material costs of this demonstrator was about 17 €(frame: 10 €, LED, photodiode: 1 €, flow cell: 1,5 €, driver electronics: 1.5 €, battery: 3 €). For adjustment purposes the polarization filters were mounted in a rotatable holder. Before the initial use of this sensor, the polarization filters are crossed and oriented as pointed out in section 4.3.2 at 45° regarding the groove lines of the PCS.

6.3.2 Detection limit in a bulk refractive index measurement

The detection limit of this demonstrator is a function of many parameters. The most important parameter might be the PCS's sensitivity, which is approx. 25 nm/RIU for the PCS used here. However, parameters such as the slope of the LED spectrum or the readout noise have a limiting effect as well. The smallest variation in bulk refractive index, which can be detected by the demonstrator is an indicator of the detection limit.

[4]Solidworks, *Dassault Systemes SolidWorks Corporation*
[5]3D Printer *Dimension*

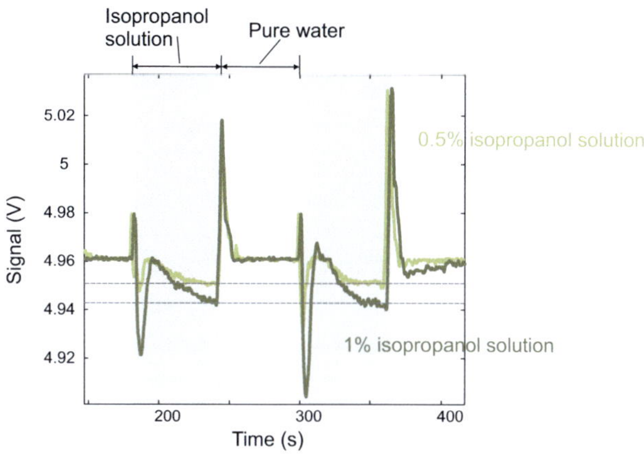

Figure 6.9: *Determination of the detection limit of the demonstrator utilizing bulk refractive index tests. A 0.5 % isopropanol-water solution could be clearly differentiated from pure water. Hence a bulk refractive index detection limit of $\Delta n = 2.23\ 10^{-4}$ is obtained.*

Three solutions are prepared with refractive indices close to each other using a water-isopropanol solution. With isopropanol concentration of 0%, 0.5% and 1% we achive a tuning of the refractive index from 1.3330 (water) to 1.333223 and 1.333446.[6] This corresponds to a Δn = 2.23 10^{-4}.

These solutions are pumped into the flow cell in two experiments. First we alternate between pure water and 1% isopropanol solution and exchange the whole volume of the flow cell in periods with a duration of 60 s. In a second experiment we alternate between water and 0.5% isopropanol. The injection process into the flow cell is performed with a syringe with a capacity of 3 ml and has a duration of about 10 s. Both results are plotted in Fig. 6.9 as the output signal in voltage as a function of time. We observe three signal levels, namely for pure water, for 0.5% and 1% isopropanol solution. These signal levels are equally spaced with a distance of about 0.13%. This result was expected, since the refractive index change as well as the response of the system are linear.[7] This experiment shows the ability of the sensor to detect a bulk refractive index change below 2.23 10^{-4}.

The reader may have noticed the signal peaks at every injection procedure. More precisely, we observe mainly a negative peak for the injection of the water-isopropanol solution and a positive peak for the injection of pure water. During these experiments we did

[6] With a refractive index of 1.3776 @20° for isopropanol.

[7] The linear response of the detection system is only valid for small refractive index changes. In these cases the GMR shift is approximately linear, since its field distribution does not change for small refractive index changes and the falling edge of the LED spectrum can be assumed linear, too.

not observe any reactions of the signal level to vibrations or displacement of the demonstrator. Thus we believe that these peak behavior is caused due to overpressure in the flow cell during the injection procedure. These peaks are either due to refractive index changes caused by changed pressure or more likely because of displacement of the PCS relative to the light source.[8]
With a closer look on the signal behavior, we observe differences after the injection of pure water and water-isopropanol solutions. The signal after pure water injection reaches quickly its final level. On the other hand, the signal after water-isopropanol solution injection shows an exponential behavior until it reaches its final level. In the scope of this work this behavior could not be investigated in more detail. However, it shows that the sensitivity of this demonstrator might be much higher than detecting a $\Delta n = 2.23\ 10^{-4}$.

6.4 Protein-protein coupling in real-time

In section 6.3.2 we showed how the proposed method can be utilized to determine a refractive index change. Such refractive index determinations are applied, for instance, in concentration measurements, such as the sugar concentration during vinification or honey production. This sensing method, however, is an ideal candidate for biosensing applications as well. A thin layer of biological substances immobilized on the surface of PCSs, cause a change of the effective refractive index of QGMs, which results in a shift of GMRs.

To validate this sensors ability as a biomolecule detector, we studied a key-lock system composed of streptavidin and biotin, two molecules that have large affinities to each other [128–131]. This affinity is among the strongest non-covalent interactions known in nature.

6.4.1 Surface functionalization with Biotin

There exist various ways to *functionalize* the PCS surface with biotin. For example a silanization process can bind biotin to a dielectric surface. This binding, however, is so strong that it cannot be removed with simple cleaning steps. Since we intend to use the same PCS for a series of experiments this kind of functionalization was not advantageous. To functionalize the PCS surfaces with biotin we chose biotinylated phospholipids, which build in a self-organizing way a bilayer of molecules with the biotin part being

[8]Because of mode dispersion in k-space any angle change results in spectral shift of mode resonances.

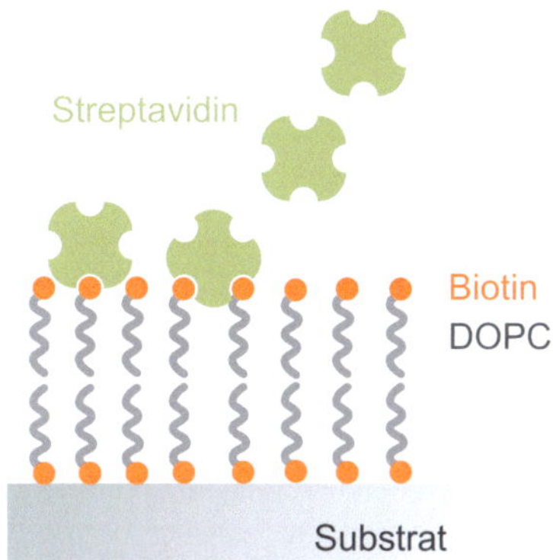

Figure 6.10: *Sketch of the binding process of streptavidin to a surface functionalized with biotin (N-Biotinyl Cap-PE). Streptavidin is a tetramer, where each subunit binds to biotin. As a result of the binding process the effective refractive index of modes in the PCS will change and influence their spectral position.*

free for binding processes (Fig. 6.10). We refer to the following literature for more details [132, 133]. A compound of biotinylated phospholipid, in particular 1,2-dioleoyl-sn-glycero-3-phosphoethanolamine-N-(cap biotinyl) (N-Biotinyl Cap-PE) mixed in mol% ratios with phospholipid 1,2-dioleoyl-sn-glycero-3-phosphocholine (DOPC) was used in order to control the amount of the functional biotin-headgroup lipids available on the surface.

We chose spin-coating as the method to apply biotinylated phospholipids on the PCS surface due to its simplicity of application and possibility to control the coating conditions. A simple but effective quality inspection was a visual test, which was the transmission image using crossed polarization filters. An inhomogeneous or too thick layer changes the transmission color of the PCS, which could be observed easily.

The N-Biotinyl Cap-PE/DOPC solution was dissolved in chloroform (20mg/ml). A further solution of 1:18 with chloroform was found to be a suitable concentration for the spin-coating process. All surfaces were spin-coated with 8000 rpm (revolutions per minute) for 45s.

6.4.2 Label-free streptavidin detection in real-time

To provide favorable conditions for the binding process of streptavdin to the biotin, we used as buffer solution phosphate buffered saline (PBS) to maintain a constant pH value of 7.4.

As the first experiment we analyzed the binding process via a spectrometer, while using an LED as the light source. In this experiment, we have injected 50 nMol of streptavidin into the flow cell and characterized the second resonance of the PCS over a total time of 1,000 seconds. The time-dependent transmission of this GMR is shown in Fig. 6.11. We observed a resonance shift of 0.5 nm as well as an intensity decrease of the transmission, which follows the falling edge of the LED spectrum.

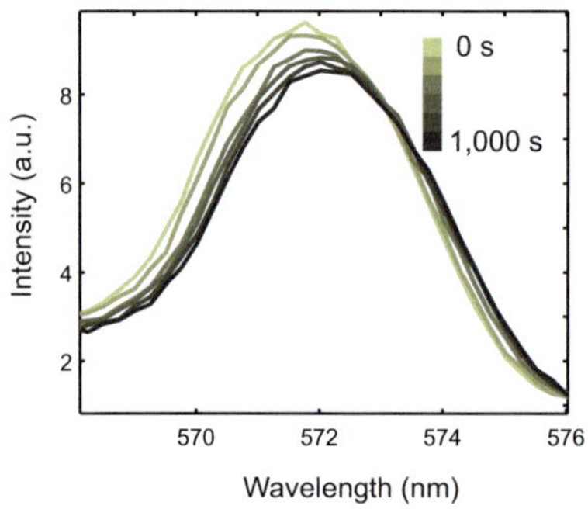

Figure 6.11: *Spectral analysis of the second resonance with a spectrometer and an LED as light source during the streptavidin coupling process. We observe a resonance shift of 0.5 nm as well as an intensity decrease of the transmission following the falling edge of the LED spectrum.*

All further experiments were performed using the demonstrator introduced in section 6.3. Fig. 6.12 shows the effect of the ratio of biotin-DOPC mixtures on the signal. The green and the red curves show the relative signal response to 25 nMol streptavidin for the surface functionalized with 10 mol% and 4 mol% biotinylated phospholipids, respectively. As expected, the streptavidin-biotin binding process was accelerated at a higher concentration of biotin. Furthermore, we observed a saturation of the signal after 1,000 s for the surface functionalized with 10 mol% biotinylated DOPC. This indicates that all the streptavidin molecules in the flow cell are bound to the surface.
The inhomogeneous decrease of the signal may be caused by the non-homogeneous functionalization, which is caused by the imperfect spin-coating process.

Using 10 mol% biotinylated phospholipids for functionalization, we reduced the concentration of streptavidin to 2.5 nMol. At this concentration we still observed a relative

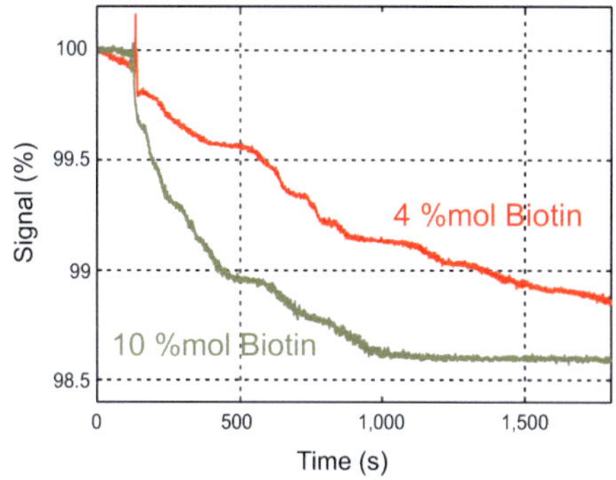

Figure 6.12: *Relative intensity reduction of GMRs as a function of time with two different composition ratios of N-Biotinyl Cap-PE in DOPC. Signal saturation is reached faster with a composition of 10 mol% N-Biotinyl Cap-PE in DOPC. Moreover an inhomogeneous decrease of the signal is observed, which is caused by the imperfect spin-coating process.*

signal reduction of about 0.4% (Fig. 6.13).

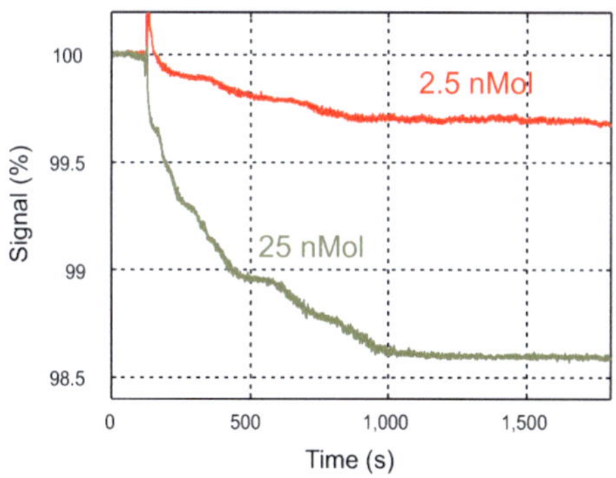

Figure 6.13: *The influence of streptavidin concentration on the signal. Lowering the streptavidin concentration the saturation level decreases too. 2.5nMol streptavidin binding to N-Biotinyl Cap-PE functionalized surface was clearly detectable.*

These graphs contain also information about the association physics of streptavidin to biotin.

This reaction can be simplified to the coupling of the ligate (streptavidin) to the immobilized ligand (biotin) on the surface of the transducer. Assuming a constant concentration of the ligate, we can describe the reaction with Langmuir's theory as [134]:

$$dR/dt = k_a C_0 \left(R_{max} - R\right) - k_d R. \tag{6.1}$$

Where R is the response of the transducer with R_{max} as the maximum response. k_a and k_d are the association and the dissociation rate constants of the ligate-ligand system, respectively. C_0 is the concentration of the ligate.
Integrating equation 6.1 we end up with an exponential curve, which is obtained by the biosensor:

$$R = \frac{C_0 k_a R_m \left(1 - e^{-(k_a C_0 + k_d)t}\right)}{C_0 k_a + k_d} + R_0. \tag{6.2}$$

Here R_0 is the initial response. Applying a fitting routine to this equation parameters such as k_a or k_d can be determined.

In this case the dissociation rate constant can be assumed to be $k_d = 0$, due to the strong coupling of streptavidin to biotin.

Chapter 7

Photonic crystal slabs as visual sensors

Summary

This chapter examines the concept of background light suppression using crossed polarization filters, which is applied to demonstrate a visual biosensor. Again, the presence of biological substances on the PCS's surface cause spectral shifts of GMRs. Here, these shifts are analyzed by the naked human eye, and are recognized as local color changes. Moreover, the ambient light is acting as the light source. Hence, this sensor functions without any power source and signal detection/processing periphery. This kind of visual sensor might be advantageous for flexible point-of-care applications.

7.1 Motivation

Today's optical sensors generally require an artificial light source, an energy supply as well as instrumentation to detect the signal. These requirements usually result in costly and non-compact devices. Moreover, these devices typically require trained technical staff.

A purely visual biosensing method requiring no power supply and being user friendly could open the door for devices applicable in many situations. In outdoor applications, for instance, the absence of a power supply is of essential value. Water tests, blood tests etc. are just some examples. On the other hand third world countries would benefit from the low costs. This would lower the threshold of adopting such devices and could improve the diagnostics of diseases.

In [22, 135] the authors introduce a visual biosensor utilizing the color change due to thin film interferences. The obtained color variations, however, exhibit low contrast and a proper sensing is difficult.

To overcome this problem a visual sensor that is based on the observation of the transmission through a PCS by the naked human eye is proposed in the following chapter. By a combination of the PCS with polarization filters the nanoscopic properties of the PCS surface are observable as color changes in ambient light. This method benefits from the fact that the human eye is more sensitive to color shifts than to intensity variations. Furthermore, as large-area PCSs can be produced cost-effectively [80, 81], this configuration is promising for future low-cost biosensors applicable without instrumentation and without prior training.

7.2 Color theory

As the sensor method introduced in this chapter is based on transmission spectra inducing a specific color to the human eye, we first discuss how to treat colors and their impression for the eye.

7.2.1 CIE 1931 color space

Colors are a human definition, as they are generated in our mind. They are nothing else but a spectral analysis of the light penetrating the eye. Each spectral distribution generates another color impression. To quantify colors different color spaces have been defined. The most common one is the CIE[1] 1931 color space, created by the International Commission on Illumination in 1931 [136].

The human eye exhibits two kinds of receptors, three different cone cells, where each of them detects a specific spectrum of the light and the rod cells, which detect only light intensity monochromatically. The response functions of the three cone cells is depicted in the *color matching functions*, which are extracted from a series of experiments with probands. In Fig. 7.1 these functions are referred as to $\overline{x}(\lambda)$, $\overline{y}(\lambda)$ and $\overline{z}(\lambda)$.

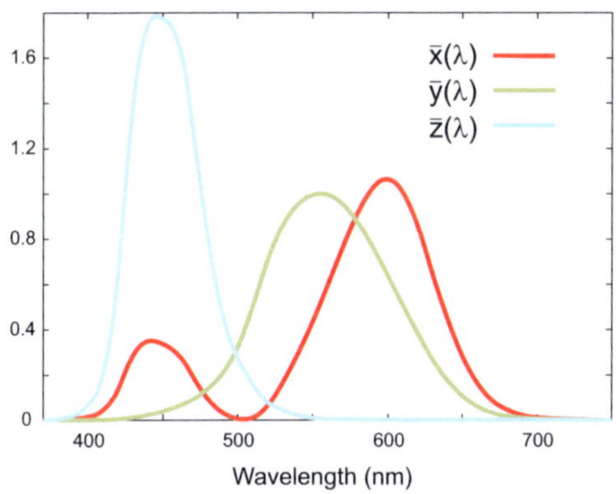

Figure 7.1: *The CIE standard observer color matching functions $\overline{x}(\lambda)$, $\overline{y}(\lambda)$ and $\overline{z}(\lambda)$. Using the spectral power distribution $I(\lambda)$ and the color matching functions the color coordinates x and y and the parameter for brightness Y can be calculated.*

To define a color the *tristimulus values* X,Y and Z, which are primary colors in a three-component additive color model, have to be determined. With the spectral power distribution $I(\lambda)$ of the spectrum of interest and the color matching functions we obtain the following terms for the tristimulus values:

$$X = \int_0^\infty I(\lambda)\,\overline{x}(\lambda)\,d\lambda \tag{7.1}$$

$$Y = \int_0^\infty I(\lambda)\,\overline{y}(\lambda)\,d\lambda \tag{7.2}$$

$$Z = \int_0^\infty I(\lambda)\,\overline{z}(\lambda)\,d\lambda \tag{7.3}$$

[1] International Commission on Illumination

With these tristimulus values one can span a three-dimensional color space. The concept of color, however, can be divided into a brightness and a color scale. For instance the colors gray and white just differ in their brightness, but not in their color. The tristimulus values are designed in such a way that Y is a measure of brightness. Hence we can reduce the color issue to only two parameters, which are a function of all tristimulus values. We define the following coordinates:

$$x = \frac{X}{X+Y+Z} \tag{7.4}$$

$$y = \frac{Y}{X+Y+Z} \tag{7.5}$$

$$z = \frac{Z}{X+Y+Z} = 1 - x - y \tag{7.6}$$

Only x and y are used to define a color in the CIE 1931 color space.

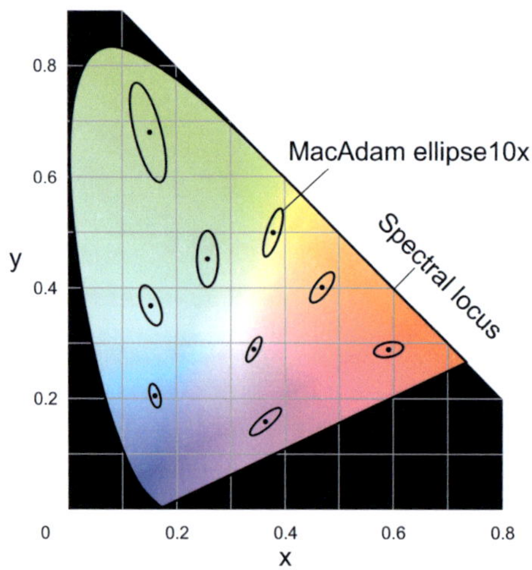

Figure 7.2: *The CIE 1931 color space chromaticity diagram, which is a mathematical system to define colors. In this color space two coordinates (x and y) define the color. An infinitely sharp resonance would be projected on the edge of the CIE color space (spectral locus), whereas an infinitely broad resonance would be positioned in the center of the CIE color space, with the coordinates x = y = 1/3. The MacAdams ellipse (here ten times bigger than its actual size) defines the region around a color point, which is indistinguishable for an average observer from the center color.*

Fig. 7.2 shows such a chromaticity diagram following the CIE 1931 color space. The outer boundary of this chromaticity diagram is the *spectral locus*. These are the coordinates of monochrome colors. The color of a monomode laser, for example, would be near to this spectral locus. On the other hand an infinitly broad resonance would be positioned in the center of the CIE color space, with the coordinates x = y = $1/3$.

7.2.2 MacAdam ellipse

For a visual sensor, which is based on color detection by human eye, a restricted color perception is a limiting factor for detecting substances. To assess this limit a series of experiments were performed in the past [137, 138]. In these experiments the probands should find their limit of distinguishing two colors placed right next to each other. The result of these experiments are the MacAdams ellipses, which define the region on a chromaticity diagram containing all colors, which are indistinguishable from the color at the center of the ellipse. In Fig. 7.2 some exemplary ellipses, being ten times larger than their actual size, are shown. We observe different sizes for these ellipses at different color points.

However, if the proband cannot compare two colors, as it was the case in these experiments, he cannot by far distinguish so many colors. This is of high value for the design of such visual sensors.

7.3 Photonic crystal slabs as visual sensors

Performing transmission experiments in the direction normal to the surface of the PCS, GMRs appear as Fano resonances superimposed with the spectrum of the light source. Because of the broadband transmission of the light source, the GMRs do not induce a specific color impression for the observer. In order to observe the GMRs, while at the same time suppressing the light not interacting with the PCS, two orthogonally oriented polarization filters are inserted before and after the PCS. As we saw in section 4.2 this configuration transmits only GMRs. For the observer the sample appears colored with the color point depending on the number and spectral position of the GMRs. Some example transmission measurements through PCSs[2] are shown in Fig. 7.3. Here the different colors are a consequence of different periodicities of the PCSs.

Any change in the optically relevant parameters, such as a change in refractive index of the surrounding medium, or the presence of a layer on the surface of the PCS, will cause either a shift in spectral position or a redistribution of mode intensities. The direct

[2]These PCSs are composed of an ITO layer with a periodic nanostructure fabricated with laser interference lithography.

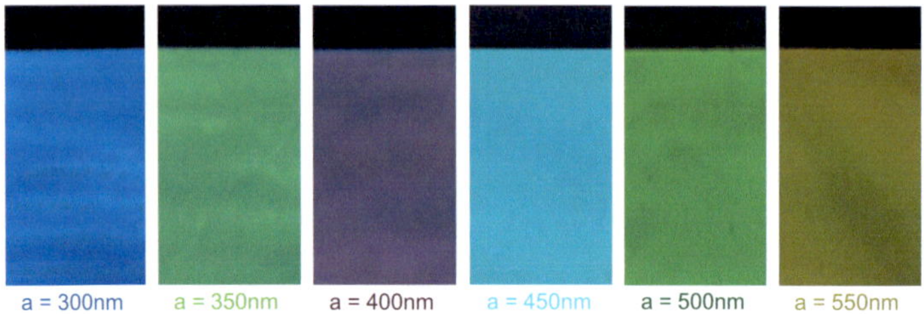

Figure 7.3: *Transmission images of PCS fabricated with laser interference lithography with crossed polarization filters. The observed color is a function of the GMRs provided by the PCSs. Here the various periodicities of the PCS cause the variety of transmission colors.*

consequence for the observer is a change in color impression. Spatially resolved surface functionalization, like immobilized antibodies [16, 139], can make the desired refractive index change on the surface of the PCS, when the biological material of interest is present. In Fig 7.4 the setup of the proposed sensor is presented. Here the light source is the ambient light. However, a light source with a low color rendering index due to sharp emission lines, such as a fluorescent lamp, might cause non predictable results.

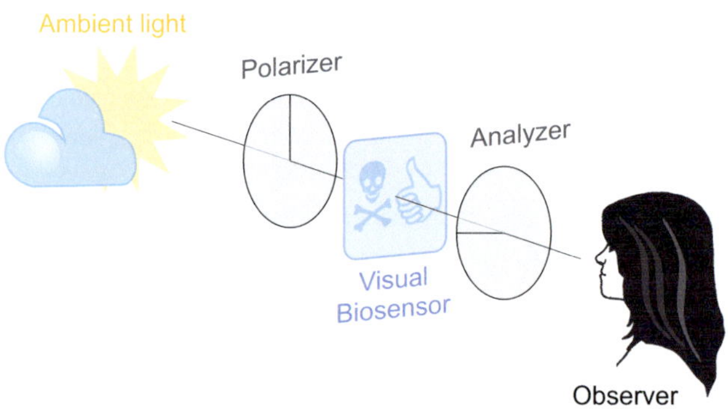

Figure 7.4: *Sketch of the setup for the visual biosensor. The light source is the ambient light, while the transmission is detected by the human eye. This configuration does not need any power supply. Therefore, it is an ideal candidate for outdoor applications.*

This sensor method leads to unlimited design possibilities, which include any symbols or characters in any color. In particular illiterate people will benefit from these design possibilities. Some fictitious examples shown in Fig. 7.5 depict the potential of this method.

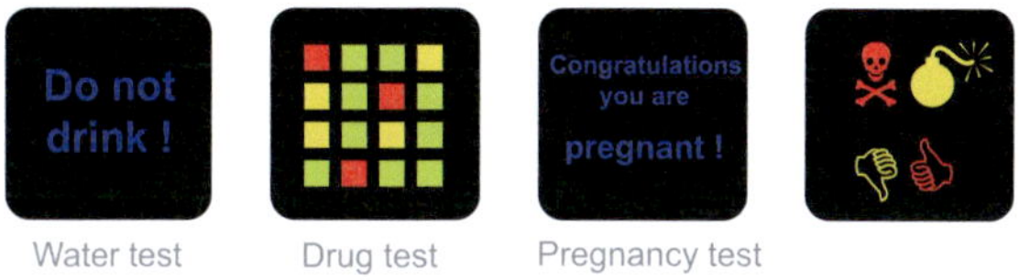

Figure 7.5: *Some fictitious examples for visual tests. The design of such visual sensors can resort to any symbols or characters in any color. In particular illiterate people will benefit from this design possibilities.*

7.3.1 Visual impression in a bulk refractive index test

To investigate this sensor method, we proceed similar to the investigations for the optoelectronic biosensor presented in chapter 6. In the first step we demonstrate the refractometric character of the proposed sensor, which is sensitivity to the refractive index of a liquid. We compare the color of the transmission through the PCS embedded in air and with three different sucrose solutions. By changing the sucrose concentration we can tune the refractive index of the solution from 1.33 (pure water) to 1.55 (1.44 g/ml).

Throughout this chapter we use a PCS consisting of holes in a Nb_2O_5 layer (140 nm) with a refractive index of 2.3 (for λ = 500 nm) on a quartz substrate in a quadratic geometry. The holes have a quadratic shape and an edge length of 380 nm. For the periodicity of the PCS we chose Λ = 500 nm, this structure shows third order resonances in the visible range, while it is easy to fabricate. To capture the visual impression of a human eye, we used the microscope setup introduced in section 4.1.1 with a 40x objective. For color images we use a single-lens reflex camera, while spectral measurements were performed in a confocal configuration.

Fig. 7.6 shows the transmission images of the PCS field surrounded with air and three sucrose solutions. The PCS appears green to the observer if surrounded by air. Using the sucrose solution with refractive indices of 1.34, 1.47 and 1.55, the observer notices a color switch from orange to red, depending on the refractive index of each solution.

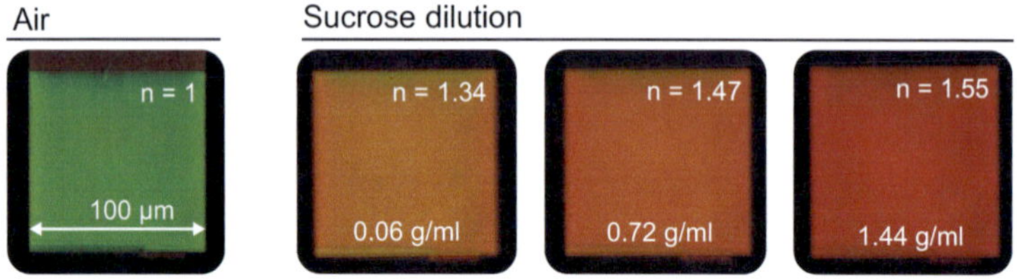

Figure 7.6: *Transmission image with crossed polarization filters of a PCS embedded in air and different liquids. The observed color for air as the surrounding material is green. Surrounded by different sucrose solutions, however, the color switches from orange to red, depending on the sucrose concentration.*

7.3.2 Guided-mode resonance shift and redistribution

To understand these results we perform spectral analysis of the transmission. For this purpose we used the confocal setup described in section 4.1.1.

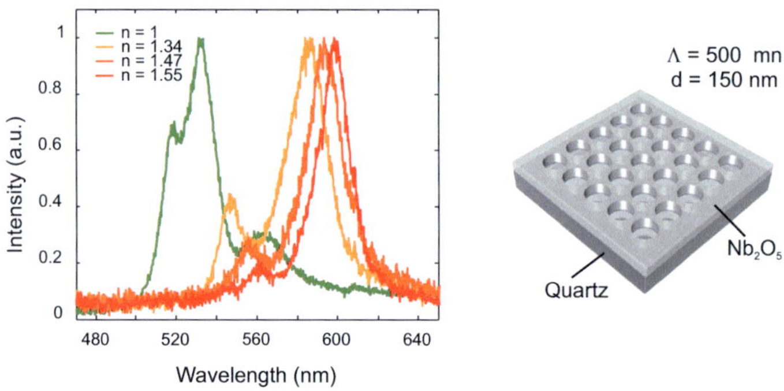

Figure 7.7: *Spectral analysis of bulk refractive index experiments shown in Fig. 7.6. Two modes are identified to contribute to the color impression. Changing the refractive index of the surrounding media results in a spectral shift of these modes.*

With air as the filling material the transmission shows two main resonances at 532 nm and 562 nm (Fig. 7.7). We refer to these resonances as mode 1 and mode 2 (Fig. 7.8). The color impression as observed in Fig. 7.6 arises from a mix of the resonance positions and intensities. Here the mode 1 contributes with a larger fraction to the color impression than the mode 2 and the observer recognizes the transmission as green. In a similar way the transmission with the sucrose solution appears orange to red.

Analyzing the transmission spectra we can identify two effects, which are responsible for the color shift, namely the resonance shift due to the Bragg condition and the intensity

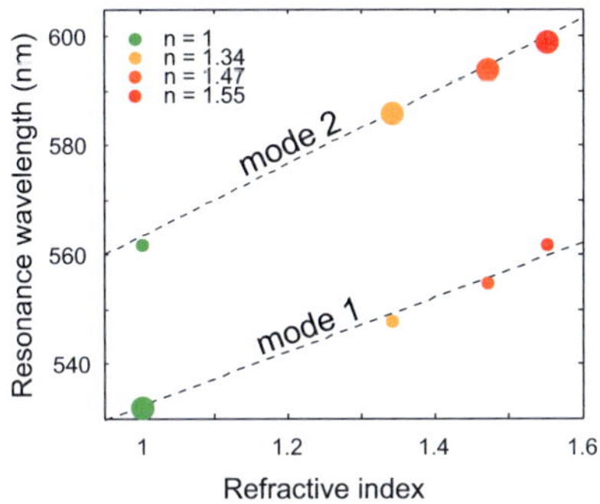

Figure 7.8: *Spectral mode shift analysis of modes shown in Fig. 7.7. Mode 1 and mode 2 shift independently with different sensitivities: 55 nm/RIU and 68 nm/RIU respectively. The dot size indicates the intensity distribution of the two modes. The dashed line is the third order Bragg condition.*

redistribution of the two modes. In Fig. 7.8 we investigate these two effects and plot both resonance maxima for each refractive index environment, where the larger dots represent the resonances with higher intensity. The resonance maxima can be derived from the Bragg condition.

$$\lambda_{res1,2} = 2\frac{\Lambda}{m}n_{eff} \tag{7.7}$$

$$n_{eff} = a_{enviro.}n_{enviro.} + a_{Nb_2O_5}n_{Nb_2O_5} + a_{sub.}n_{sub.} \tag{7.8}$$

where Λ is the photonic crystal periodicity and m is the order of the Bragg condition. n_{eff} is calculated from the sum of the contributing refractive indices and their participation to each mode. The resonance position of both modes follow the third order Bragg condition, as depicted in Fig. 7.8 (dashed lines). The different resonance positions are a consequence of the different n_{eff} for these modes. We determine sensitivities of $\Delta\lambda/\Delta$n = 55 nm/RIU and 68 nm/RIU for the 1. and 2. mode, respectively. However, the overall sensitivity of this PCS is higher, due to the mode intensity redistribution. PCSs with higher sensitivities afford the percipience of smaller refractive index changes.

To indicate the color shift progress, we plot the chromaticity in the CIE color space, as depicted in Fig. 7.9. As expected the color of the transmission of the PCS embedded in air and sugar solutions positioned on a line parallel to the spectral locus, which is the consequence of the finite width of the resonances. The human eye is not the best detector

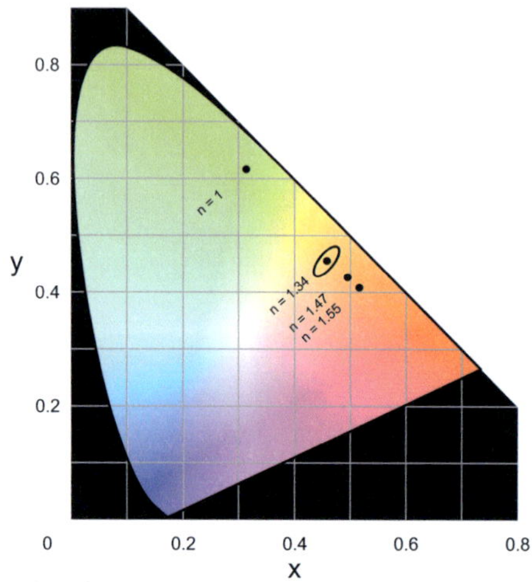

Figure 7.9: *The CIE 1931 color space chromaticity diagram with the calculated color point using the spectrum in Fig. 7.7. These point are on a line parallel to the spectral locus. The MacAdam ellipse (ten times magnified) around the color point for n = 1.34 indicates the distinguishability of all color points.*

to determine colors absolutely, but is very good in determining them relatively. In Fig. 7.9 we plot around the color point for n = 1.34 the MacAdam ellipse with 10x magnification. This is another proof that these colors are clearly distinguishable for the average observer.

7.4 Photonic crystal slab as visual biosensor

We prove the ability of the proposed biosensor to detect biological substances by infiltrating the PCS surface with human epithelial cheek cells. To place these cells on the PCS surface, we put a water solution including the cells on the surface of the PCSs. The cells deposit on the PCS surface after the water solution is dried up. The refractive index of a human epithelial cell is around 1.45, which is comparable to the refractive index of the investigated sugar solutions. In Fig. 7.10 the transmission trough the PCS with a human cheek cell on its surface, using crossed polarization filters, is shown. The transmission color of the areas in contact with the cheek cell is orange to red, which is highly consistent with the sugar solution experiments. This indicates that the holes of the PCS are filled with the cell membrane. On the other hand the cheek cell on the substrate and without crossed polarization filters, shows almost no color contrast.

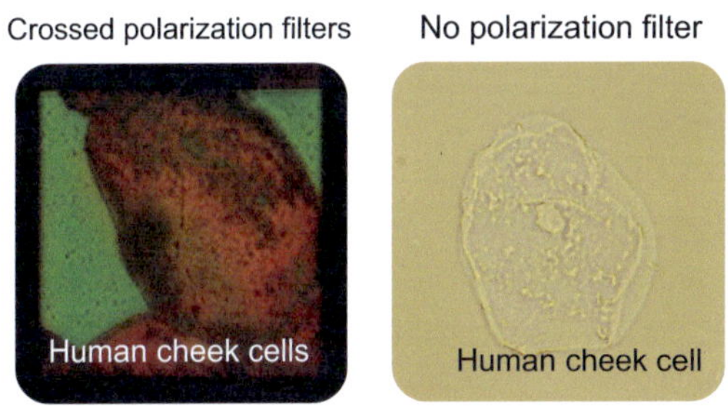

Figure 7.10: *A human epithelial cheek cell placed on the surface of the PCS with crossed polarization filters and on the substrate without crossed polarization filters. As expected, the areas covered with the cheek cell appear red, while the bare parts appear green (left picture). On the other hand, the human cheek cell is transparent on the substrate without polarization filters (right picture).*

Chapter 8

Enhanced contrast in microscopy

Summary

The microscopy of biological materials, such as cells, suffers from their transparency. To enhance the contrast methods such as phase contrast microscopy or fluorescence microscopy are applied.
In this chapter a novel method for enhanced contrast in the microscopy is presented. This method is based on a PCS as the microscope slide between two crossed polarization filters. Two techniques are presented to enhance the contrast of the specimen: enhanced contrast due to guided-mode resonance shift and enhanced contrast due to appearance of guided-modes resonances.

8.1 Motivation

The visibility of details in an image depends on the resolution and the contrast of the image. In particular, the optical microscopy of biological objects, for instance cellular specimen, is hampered by contrast poverty of the images. This is the case, whenever the objects do not appear with sufficient contrast by general light absorption (darkening), by specific spectral light absorption (intrinsic color) or by high differences in their refractive indices. Therefore, there are various methods to increase the contrast. For example, individual objects or their components are specifically marked with dyes. It is an advantage to choose colors that have a high contrast to each other, this is the case for complementary colors. The method of staining, however, shows limitations particularly in the investigation of living cells.

Another method is to reduce the illuminating aperture using the iris diaphragm of the condenser. This leads to a contrast enhancement of the object due to its varying refractive index. As the resolution of the microscope is a function of the numerical aperture of the illumination, this method has the disadvantage that the resolution of the image is greatly reduced.

A major advance in the increase of contrast was the invention of the phase contrast microscope by Frits Zernike, for which he was honored with the Nobel Prize in 1953. With the phase contrast microscope it is possible to generate an intensity contrast without substantially reducing the resolution. The phase contrast microscope uses the phase difference that results when light passes through a medium with higher refractive index than its surrounding area. Through systematic manipulation of the phase of the background light and the light passing through the object the contrast is enhanced. A so-called phase ring is used to influence the background light.

In this chapter we will introduce an alternative way to enhance the contrast in the light microscopy. We propose to use a PCS as the microscope slide. The presence of a specimen on the surface of the PCS influences its GMRs. Using crossed polarization filters the influence on the GMRs can be visualized to the observer. This method do not need any staining technique. Therefore, it is ideal for living specimen microscopy. Furthermore, this method is based on a conventional light microscope, which renders it cost-efficient.

In the following we capture the visual impression of the contrast enhancement with the setup introduced in section 4.1.1. These pictures were captured using a single lens camera. The reader should note that these pictures cannot reflect exactly the image observed by the human eye. In particular in the deep blue region the image taken with the single lens camera is not representative for the real impression.

8.2 Enhanced contrast due to guided-mode resonance shift

In this chapter we introduce two ways to enhance the contrast utilizing the PCS. First we discuss the enhancement due to GMR shift. As we already learned from chapters 6 and 7, the presence of any material with differing refractive index causes a shift of GMR's spectral position. Now if using a PCS as the microscope slide with the specimen on its surface, the GMR's spectral position will shift on regions covered with the specimen depending on its refractive index. Moreover, this shift is also a function of the thickness of the specimen. Using crossed polarization filters only GMRs are transmitted. The shift in GMR's spectral position will cause a color shift of the PCS and the specimen experience a contrast enhancement.

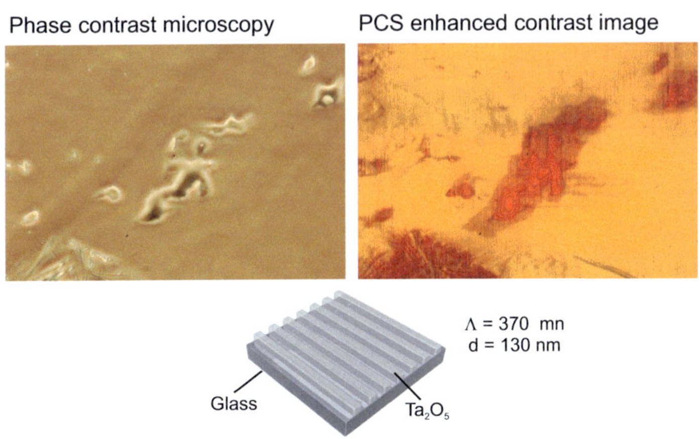

Figure 8.1: *Microscope image of N-Biotinyl Cap-PE remains on PCS after spin coating process. Left image: phase contrast microscope image. Right image: conventional light microscopy with PCS and crossed polarization filters. Color change at regions covered with N-Biotinyl Cap-PE due to GMR shift. Crossed polarization filters reveals thin films on the PCS, which were hidden using phase contrast microscopy.*

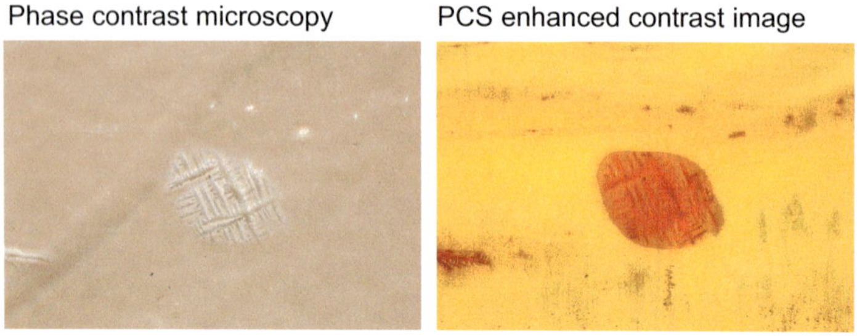

Figure 8.2: *Microscope image of N-Biotinyl Cap-PE remain after spin coating process.*

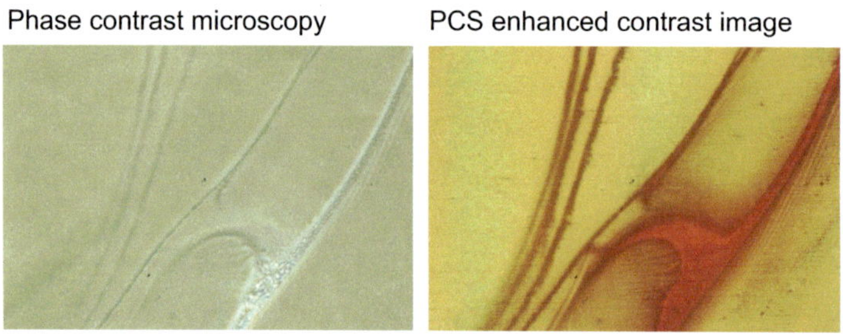

Figure 8.3: *Microscope image of N-Biotinyl Cap-PE remain after spin coating process.*

In Fig. 8.1, 8.2 and 8.3 three examples using this method are shown. Images on the left are captured in a phase contrast setup, while the images on the right side are captured with crossed polarization filters. The PCS[1] is composed of a Ta_2O_5 layer with a linear grating showing a periodicity of Λ = 370 nm. The structures on the PCSs are remains of N-Biotinyl Cap-PE during the imperfect spin coating process performed for experiments in chapter 6. These pictures, however, show in an impressive way the higher contrast using PCS as microscope slide compared to phase contrast microscopy. In particular, the contrast of very thin films on the microscope slide is much enhanced compared to phase contrast microscopy.

[1]This PCS was used for the optoelectronic biosensor introduced in chapter 6, too.

8.3 Enhanced contrast due to arising of guided-mode resonances

As a second way to enhance the contrast of specimen on a PCS as the microscope slide we study a similar phenomenon, which was already discussed in section 3.4.1, namely the arise of GMRs when changing the geometrical parameters. In section 3.4.1 GMRs disappear with a hole diameter, which is too big. This is due to the absence of the high index material in the holes, which results in a too low effective refractive index for the modes to enable mode guiding.

8.3.1 Guided-mode resonances as a function of the refractive index

Using a PCS with big hole radii, as exemplary shown in Fig. 8.4, in a transmission experiment, no GMRs are observed. In a crossed polarization configuration this would result in a black background. One way to increase the effective refractive index and induce GMRs is to increase the refractive index of the surrounding material, for example replacing the surrounding air with a liquid.

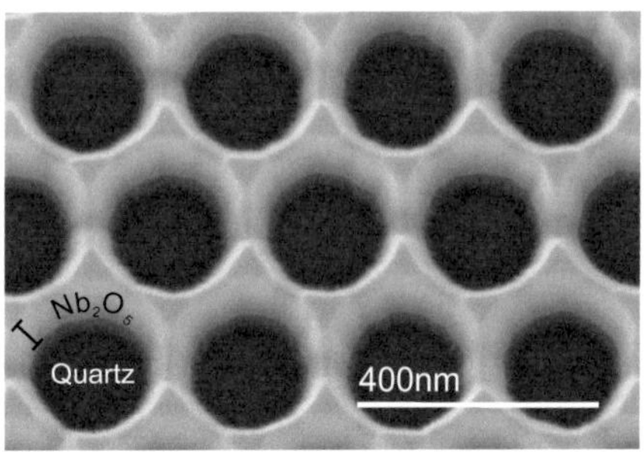

Figure 8.4: *SEM image of an exemplary PCS with big hole diameter. Due to the absence of the high index layer in the hole, the effective refractive index becomes too low, to guide modes in the high index layer. As a consequence no GMRs are obtained in the transmission measurements.*

We do so and prepare two sucrose solutions with 0.06 g/ml and 1.44 g/ml, respectively. In Fig. 8.5 transmission images through a PCS composed of a N_2O_5 layer on a quartz substrate with holes in a triangular geometry are shown. These images appear blue due to the spectral position of GMRs provided by this structure. As expected we observe a relation between the refractive index of the surrounding medium and the intensity of transmitted GMRs. The higher the refractive index the higher the transmission intensity. This phenomenon converts the refractive index change on the surface of the PCS into an intensity change. Hence, it can be applied to the microscopy of specimen with a different refractive index to their surroundings.

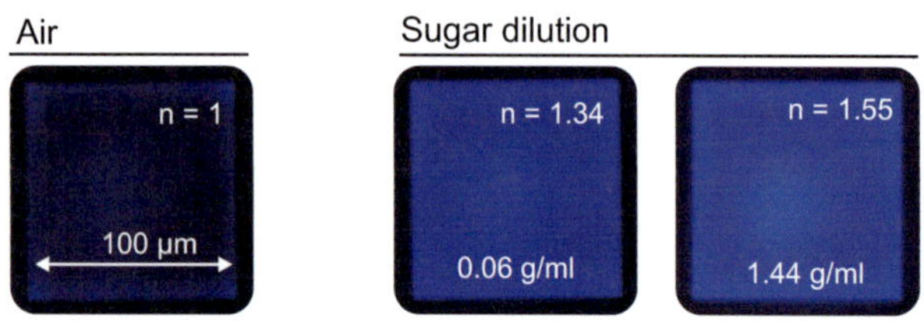

Figure 8.5: *Transmission images of a PCS infiltrated with media possessing different refractive index. The higher the refractive index of the surrounding medium the higher the transmission of GMRs.*

To show the ability of this method for contrast enhancement, we infiltrate the PCS surface with human epithelial cheek cells. For this purpose a water solution including the cells was put on the PCS surface. This solution dried out and cells were deposited on the surface. In Fig. 8.6 we compare the contrast obtained in a conventional light microscope with the method proposed here. These two images were captured at two different illumination intensities. For images with crossed polarization filters we need a higher illumination intensity, as only GMRs are transmitted, which are spectrally limited. Two human cheek cells are on the sample. Cell 1 is on an area without nanostructuring, while cell 2 is on a PCS field. The contrast using the light microscope is very poor for both cells (left image). On the other hand using crossed polarization filters cell 2, which lies on the PCS surface experiences a contrast enhancement, while cell 1 disappears into the dark background. With a closer look on cell 2 we observe a higher intensity transmission for its cell nucleus. This might be caused due to its higher refractive index.

In another series of experiments we compare the contrast enhancement with phase contrast microscopy. In these experiments we use lung carcinoma cells, which where grown on the sample in an incubator.[2] In Fig. 8.7 three microscopy methods are compared: conventional light microscopy, phase contrast microscopy and light microscopy with crossed polarization filters. With the conventional light microscopy almost no contrast

[2]These cells are epithelium cells and therefore grow favored on surfaces.

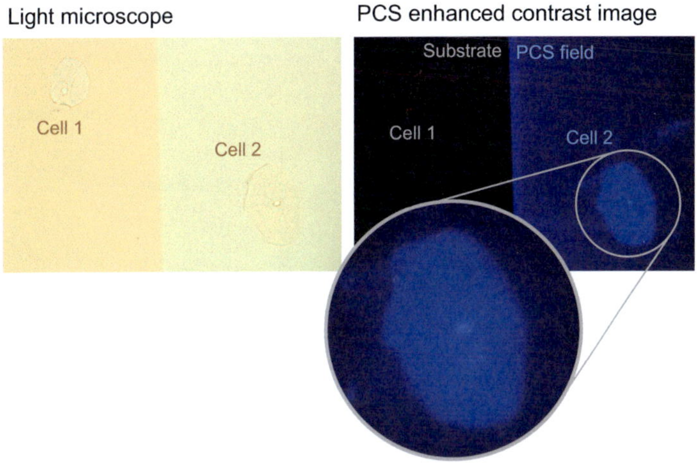

Figure 8.6: *Microscope image of 2 human epithelial cheek cells under light microscopy with and without crossed polarization filters. Without polarization filters almost no contrast is observed. Using polarization filters, however, causes a modulation in the intensity and the cheek cell on the PCS field experiences a contrast enhancement. Moreover, parts of the cell with a higher refractive index like its nucleus, appear brighter.*

is observed. Using phase contrast microscopy the contrast of these cells is increased. However, it is difficult to determine the border of each cell and its adhesion area to the substrate. Applying crossed polarization filters only the contrast of objects close to the surface is enhanced and we observe an image of cell parts adhering to the surface with enhanced contrast.

The contrast enhancements shown in this work were a modulation of the intensity of the color blue. However, tuning the periodicity of the used PCS, this intensity modulation can be applied to other colors,[3] too.

[3]Which lies on the spectral locus of the CIE color space.

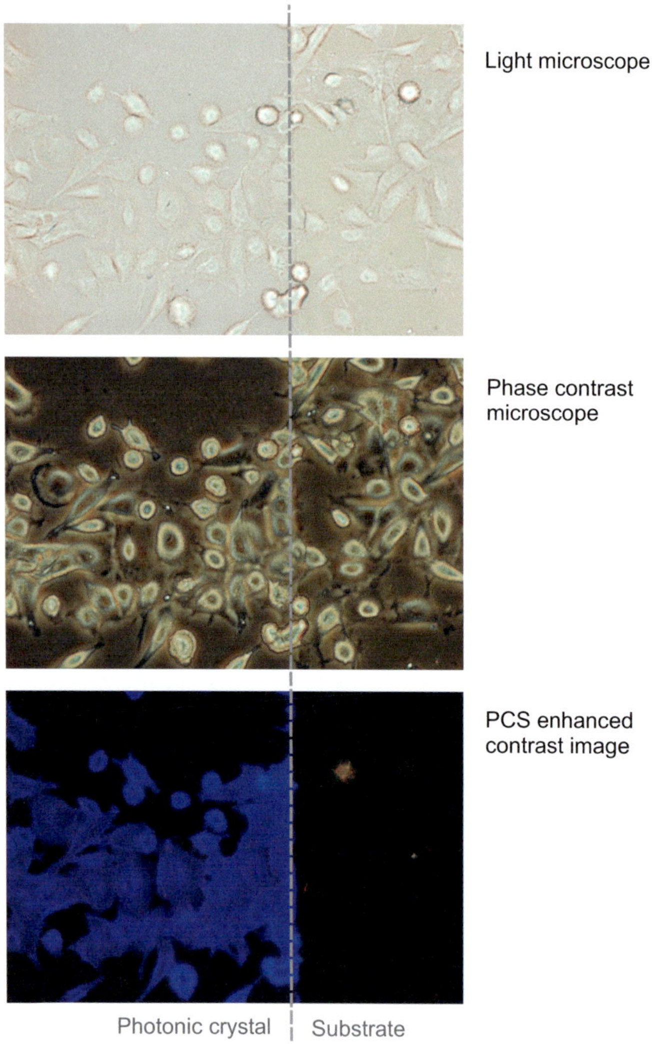

Figure 8.7: *The comparison of three microscopy methods: conventional light microscopy, phase contrast microscopy and PCS enhanced contrast microscopy.*

8.3.2 Verification using FDTD simulations

To verify the assumption of the GMR arising due to a refractive index change, we perform FDTD transmission simulations. Before doing so, we have to determine the geometrical parameters of the PCS. We proceed as described in section 5.2 and obtain the following parameters: slab thickness was d = 221 nm, periodicity Λ = 280 nm, hole radius r =140 nm. Furthermore we assumed a non-cylindrical hole profile, which follows a polynomial fifth degree.

The comparison of the transmission obtained by simulations and in the experiment surrounded by water is shown in Fig. 8.8.

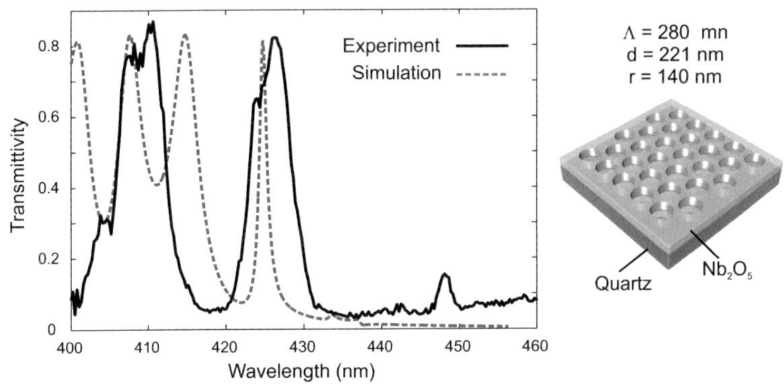

Figure 8.8: *Transmittivity through the PCS surrounded by water obtained by experiment and FDTD simulation.*

Using the determined geometrical parameters we perform a series of transmission simulations, while tuning the refractive index of the surrounding material (air (n = 1) till cellular tissue (n = ca. 1.5)). Fig. 8.9 shows the resulting GMRs for s and p polarization. These simulations reveal two effect. The GMR's shift as obtained in many situations in this thesis and as predicted GMR's arise with increasing refractive index.

8.3.3 Spatial resolution of guided-mode resonances

In these investigations the spatial resolution gained by this system is not limited by the optics used for the magnification, but is a function of the propagation length of QGM in the slabs structure, which is sensing the biological tissue. The propagation length of

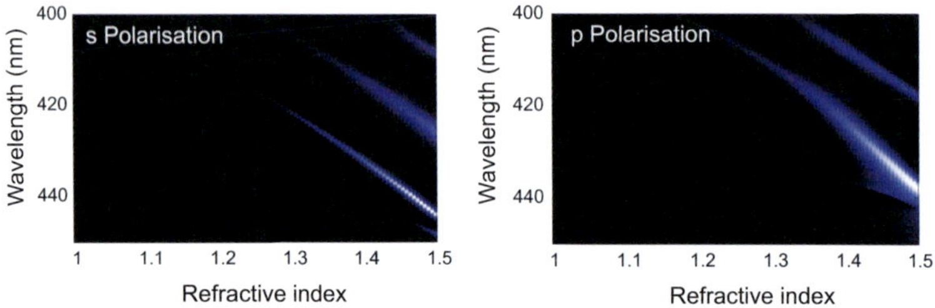

Figure 8.9: *Transmittivity through the PCS as a function of the refractive index of the surrounding medium obtained by FDTD simulations. We observe a GMR shift, as well as the appearance of GMRs with increased refractive index.*

QGMs is related strongly to the duration of the photons in the structure, which is its lifetime τ_p. If neglecting any absorption of the electromagnetic waves, τ_p can be written as:

$$\tau_p = \frac{Q}{\omega_0} = \frac{Q\lambda_0 n_{eff}}{2\pi c}. \tag{8.1}$$

Here Q is the Q-factor of the contributing mode and λ_0 its central wavelength, c is the speed of light and n_{eff} the effective refractive index of the mode. This definition for τ_p is the time, in which the photon is diffracted to the far field with a probability of 1-1/e. To determine the propagation length, which is the distance where 1/e photons of the resonance remain, we apply the lifetime of the photons and end up with the following equation:

$$L_p = \tau_p \frac{c}{n_{eff}} = \frac{Q\lambda_0}{\pi}. \tag{8.2}$$

From this equation obtain out that the lower the Q-factor of contributed QGMs the higher the resolution. In this work we analyzed modes (Fig. 5.22, p. 90) with a Q-factor down to 50 with the central wavelength at 530 nm. This yields a resolution of about 8.5 μm. In [140], however, the authors claim a propagation length down to sub micron regime.

Chapter 9

Conclusions and outlook

In conclusion, this work introduces an easy-to-implement technology platform for biosensors based on functionalized photonic crystal slabs in combination with crossed polarization filters, which was presented in chapter 4. The presence of the polarization filters allows for a direct access to GMRs of the PCS, which results in various design possibilities for biosensors. These sensors can be fabricated very cost-efficiently and be used to design user-friendly devices. Based on the proposed platform, we presented two biosensor types. In chapter 6, an optoelectronic biosensor that uses an LED as light source and a photodiode as detector to perform affinity measurements in real time, was described. This device is capable to detect a 2.5 nMol streptavidin solution in PBS, which is a relevant concentration in life sciences. The second sensor type, as introduced in chapter 7, is a visual biosensor based on the observation of a surface functionalized PCS's color embedded between two crossed polarization filters. A refractive index change on the surface of the PCS resulted in the experiment where a human epithelial cheek cell generated a directly apparent color shift. Sensors using larger PCSs, potentially allow naked-eye inspection of specific bacterial or viral materials. This sensor functions without any power source; therefore, it is ideal for cost-efficient and flexible point-of-care applications.

Moreover, in this work a novel technology for contrast enhancement in microscopy is introduced. This technology is based on a PCS placed between two crossed polarization filters acting as the microscope slide. Tissues in direct contact with the PCS experience a contrast enhancement. Two effects were found to be crucial for the contrast enhancement, namely GMR shift and GMR appearance.

To guarantee sensitive and correctly working biological applications using PCS, proper quality control is an essential task. This issue was discussed in chapter 5, where a visual and a spectrometer based inspection method were introduced. In the scope of this thesis a microscope setup was constructed, which was capable for spatially and angle resolved transmission measurements. Furthermore, this setup was used for spatially and angle resolved measurements of GMR's Q-factor.

9.1 Optimization of photonic crystal slabs for sensing applications

The design of PCSs with the proper material choice and its geometrical arrangement is crucial for photonic crystal mode field distribution. This field distribution on the other hand is strongly related to the sensitivity of the desired sensor. The simulation method introduced in chapter 3 is a perfect tool for optimizing the design of PCSs in terms of their sensitivity. The intensity filed distribution of modes, provided by PCSs, can be analyzed to determine the fraction of the mode penetrating out of the structure. This part of the mode can interact with changes on the surface of the PCS and changes its effective refractive index n_{eff}. Modes in a PCS can be classified into two groups: *dielectric modes* and *air modes*. Dielectric modes show their intensity peaks in the dielectric, while air modes have their intensity peak outside the dielectric. Hence, air modes are more suitable for sensing applications.

In the scope of this work, fabrication processes were limited to few PCS designs. Therefore, the sensitivities of presented applications suffer from these limitations. In future work the optimization of PCS should be carried out.

9.2 Future sensor applications

The introduced technology platform can be used to realize a variety of cost-efficient sensors. Non-biological sensing applications, such as temperature or pressure sensors, can be realized with the same technology and can be integrated on the same biochip to monitor the environmental conditions of experiments.

Other types of biosensors, e.g., imaging biosensors [141, 142], may also be realized using the same technology platform and would equally benefit from the efficient background suppression.

Bibliography

[1] E. Yablonovitch.
Inhibited spontaneous emission in solid-state physics and electronics.
Physical Review Letters, 58:2059–2062, 1987.

[2] S. John.
Strong localization of photons in certain disordered dielectric superlattices.
Physical Review Letters, 58:2486–2489, 1987.

[3] J. W. S. Rayleigh.
On the remarkable phenomenon of crystalline reflexion described by Prof. Stokes.
Philosophical Magazine, 26:256 – 265, 1888.

[4] K. Iga.
Surface-emitting laser - Its birth and generation of new optoelectronics field.
IEEE Journal on Selected Topics in Quantum Electronics, 6:1201 – 1215, 2000.

[5] T. Baba.
Slow light in photonic crystals.
Nature Photonics, 2:465 – 473, 2008.

[6] H. Kosaka, T. Kawashima, A. Tomita, M. Notomi, T. Tamamura, T. Sato, and S. Kawakami.
Superprism phenomena in photonic crystals.
Physical Review B, 58:10096 – 10099, 1998.

[7] B. E. Nelson, M. Gerken, D. A. B. Miller, R. Piestun, C. C. Lin, and J. S. Harris.
Use of a dielectric stack as a one-dimensional photonic crystal for wavelength demultiplexing by beam shifting.
Optics Letters, 25:1502 – 1504, 2000.

[8] B. T. Cunningham, P. Li, S. Schulz, B. Lin, C. Baird, J. Gerstenmaier, C. Genick, F. Wang, E. Fine, and L. Laing.
Label-free assays on the bind system.
Journal of Biomolecular Screening, 9:481, 2004.

[9] R. Magnusson and S. S. Wang.
New principle for optical filters.
Applied Physics Letters, 61:1022 – 1024, 1992.

[10] S. Fan and J. D. Joannopoulos.
Analysis of guided resonances in photonic crystal slabs.
Physical Review B, 65:235112, 2002.

[11] A. Ymeti, J. Greve, P. Lambeck, T. Wink, S. van Hövell, T. A. M. Beumer, R. R. Wijn, R. G. Heideman, V. Subramaniam, and J. S. Kanger.
Fast, ultrasensitive virus detection using a young interferometer sensor.
Nano Letters, 7:394 – 397, 2007.

[12] M. F. Pineda, L. L. Chan, T. Kuhlenschmidt, C. J. Choi, M. Kuhlenschmidt, and B. T. Cunningham.
Rapid specific and label-free detection of porcine rotavirus using photonic crystal biosensors.
IEEE Sensors Journal, 9:470, 2009.

[13] G. Zheng, F. Patolsky, Y. Cui, W. U. Wang, and C. M. Lieber.
Multiplexed electrical detection of cancer markers with nanowire sensor arrays.
Nature Biotechnology, 23:1294 – 1301, 2005.

[14] D. R. Thévenot, K. Toth, R.A. Durst, and G.S. Wilson.
Electrochemical biosensors: Recommended definitions and classification.
Pure and Applied Chemistry, 71:2333 – 2348, 1999.

[15] F. W Scheller, U. Wollenbergera, A. Warsinkea, and F. Lisdata.
Research and development in biosensors.
Current Opinion in Biotechnology, 12:35 – 40, 2001.

[16] L. Nicu and T. Leïchlé.
Biosensors and tools for surface functionalization from the macro- to the nanoscale: The way forward.
Journal of Applied Physics, 104:111101, 2008.

[17] D. Ivnitski, I. Abdel-Hamid, P. Atanasov, and E. Wilkins.
Biosensors for detection of pathogenic bacteria.
Biosensors and Bioelectronics, 14:599 – 624, 1999.

[18] Y. Cui, Q. Wei, H. Park, and C. M. Lieber.
Nanowire nanosensors for highly sensitive and selective detection of biological and chemical species.
Science, 293:1289 – 1292, 2001.

[19] J. S. Daniels and N. Pourmand.
Label-free impedance biosensors: Opportunities and challenges.
Electroanalysis, 19:1239, 2007.

[20] M. Wiki and R. E. Kunz.
Wavelength-interrogated optical sensor for biochemical applications.
Optics Letters, 25:463 – 465, 2000.

[21] V. S.-Y. Lin, K. Motesharei, K.-P. S. Dancil, M. J. Sailor, and M. R. Ghadiri.
A porous silicon-based optical interferometric biosensor.
Science, 278:840 – 843, 1997.

[22] R. Jenison, S. Yang, A. Haeberli, and B. Polisky.
Interference-based detection of nucleic acid targets on optically coated silicon.
Nature Biotechnology, 19:62, 2001.

[23] F. Vollmer, D. Braun, A. Libchaber, M. Khoshsima, I. Teraoka, and S. Arnold.
Protein detection by optical shift of a resonant microcavity.
Applied Physics Letters, 80:4057 – 4059, 2002.

[24] C. Monat, P. Domachuk, and B. J. Eggleton.
Integrated optofluidics: A new river of light.
Nature Photonics, 1:106, 2007.

[25] A. M. Armani, R. P. Kulkarni, S. E. Fraser, R. C. Flagan, and K. J. Vahala.
Label-free, single-molecule detection with optical microcavities.
Science, 317:783 – 787, 2007.

[26] K. Schmitt, B. Schirmer, C. Hoffmann, A. Brandenburg, and P. Meyrueis.
Interferometric biosensor based on planar optical waveguide sensor chips for label-free detection of surface bound bioreactions.
Biosensors and Bioelectronics, 22:2591 – 2597, 2007.

[27] J. N. Anker, W. P. Hall, O. Lyandres, N. C. Shah, J. Zhao, and R. P. Van Duyne.
Biosensing with plasmonic nanosensors.
Nature Materials, 7:442 – 453, 2008.

[28] M. Lu, S. S. Choi, U. Irfan, and B. T. Cunningham.
Plastic distributed feedback laser biosensor.
Applied Physics Letters, 93:111113, 2008.

[29] S. Rodriguez-Mozaz, M. J. Lopez de Alda, and D. Barceló.
Biosensors as useful tools for environmental analysis and monitoring.
Analytical and Bioanalytical Chemistry, 386:1025 – 1041, 2006.

[30] M. Nayak, A. Kotian, S. Marathe, and D. Chakravortty.
Detection of microorganisms using biosensorsa smarter way towards detection techniques.
Biosensors and Bioelectronics, 25:661 – 667, 2009.

[31] J. M. Perkel.
Who needs labels?: Macromolecular interaction sans labels.
Science, pages 1561 – 1565, 2009.

[32] R. Karlsson and R. Ståhlberg.
Surface plasmon resonance detection and multispot sensing for direct monitoring of interactions involving low-molecular-weight analytes and for determination of low affinities.
Analytical Biochemistry, 228:274 – 280, 1995.

[33] J. Homola, S. S. Yee, and G. Gauglitz.
Surface plasmon resonance sensors: review.
Sensors and Actuators B, 54:3 – 15, 1999.

[34] A. J. Haes and R. P. Van Duyne.
A nanoscale optical biosensor: Sensitivity and selectivity of an approach based on the localized surface plasmon resonance spectroscopy of triangular silver nanoparticles.
Journal of the American Chemical Society, 124:10596 – 10604, 2002.

[35] N. Nath and A. Chilkoti.
A colorimetric gold nanoparticle sensor to interrogate biomolecular interactions in real time on a surface.
Analytical Chemistry, 74:504 – 509, 2002.

[36] T. M. Chinowsky, J. G. Quinn, D. U. Bartholomew, R. Kaiser, and J. L. Elkind.
Performance of the spreeta 2000 integrated surface plasmon resonance affinity sensor.
Sensors and Actuators B, 91:266 – 274, 2003.

[37] J. Homola.
Present and future of surface plasmon resonance biosensors.
Analytical and Bioanalytical Chemistry, 377:528 – 539, 2003.

[38] N. Nath and A. Chilkoti.
Label-free biosensing by surface plasmon resonance of nanoparticles on glass: Optimization of nanoparticle size.
Analytical Chemistry, 76:5370 – 5378, 2004.

[39] E. M. Purcell.
Spontaneous emission probabilities at radio frequencies.
Physical Review, 69:681, 1946.

[40] S. P. Ogawa, M. Imada, S. Yoshimoto, M. Okano, and S. Noda.
Control of light emission by 3d photonic crystals.
Science, 305:227–229, 2004.

[41] P. Lodahl, A. F. van Driel, I. S. Nikolaev nad A. Irman, K. Overgaag, D. L. Vanmaekelbergh, and W. L. Vos.
Controlling the dynamics of spontaneous emission from quantum dots by photonic crystals.
Nature, 430:654–657, 2004.

[42] M. Fujita, S. Takahashi, Y. Tanaka, T. Asano, and S. Noda.
Simultaneous inhibition and redistribution of spontaneous light emission in photonic crystals.
Science, 308:1296–1298, 2005.

[43] D. S. Wiersma, P. Bartolini, A. Lagendijk, and R. Righini.
Localization of light in a disordered medium.
Nature, 390:671–679, 1997.

[44] P. Russell.
Photonic crystal fibers.
Science, 299:358–362, 2003.

[45] J. M. Dudley, G. Genty, and S. Coen.
Supercontinuum generation in photonic crystal fiber.
Review of Modern Physics, 78:1135–1184, 2006.

[46] E. Yablonovitch, T. J. Gmitter, and K. M. Leung.
Photonic band structure: The face-centered-cubic case employing nonspherical atoms.
Physical Review Letters, 67:2295–2298, 1991.

[47] A. A. Zakhidov, R. H. Baughman, Z. Iqbal, C. Cui, I. Khayrullin, S. O. Dantas, J. Marti, and V. G. Ralchenko.
Carbon structures with three-dimensional periodicity at optical wavelengths.
Science, 282:897 –901, 1998.

[48] M. Hermatschweiler, A. Ledermann, G. A. Ozin, M. Wegener, and G. von Freymann.
Fabrication of silicon inverse woodpile photonic crystals.

Advanced Materials, 17:2273 – 2277, 2007.

[49] P. R. Villeneuve, S. Fan, S. G. Johnson, and J. D. Joannopoulos.
Three-dimensional photon confinement in photonic crystals of low-dimensional periodicity.
IEEE Proceedings-Optoelectronics, 145:384 – 390, 1998.

[50] E. Chow, S. Y. Lin, S. G. Johnson, P. R. Villeneuve, J. D. Joannopoulos, J. R. Wendt, G. A. Vawter, W. Zubrzycki, H. Hou, and A. Alleman.
Three-dimensional control of light in a two-dimensional photonic crystal slab.
Nature, 407:983 – 986, 2000.

[51] S. H. Fan, P. R. Villeneuve, J. D. Joannopoulos, and E. F. Schubert.
High extraction efficiency of spontaneous emission from slabs of photonic crystals.
Physical Review Letters, 78:3294–3297, 1997.

[52] J. D. Joannopoulos, R. D. Meade, and J. N. Winn.
Photonic Crystals: Molding the Flow of Light.
Princeton Academic Press, New Jersey, 1995.

[53] K. M. Ho, C. T. Chan, and C. M. Soukoulis.
Existence of a photonic gap in periodic dielectric structures.
Physical Review Letters, 65:3152–3155, 1990.

[54] R. D. Meade, A. M. Rappe, K. D. Brommer, J. D. Joannopoulos, and O. L. Alerhand.
Accurate theoretical analysis of photonic band-gap materials.
Physical Review B, 48:8434–8437, 1993.

[55] L. C. Andreani and D. Gerace.
Photonic-crystal slabs with a triangular lattice of triangular holes investigated using a guided-mode expansion method.
Physical Review B, 73:235114, 2006.

[56] J. D. Joannopoulos, S. G. Johnson, and J. N. Winn.
Photonic crystals: molding the flow of light.
Princeton University Press, 2008.

[57] S. S. Wang, R. Magnusson, J. S. Bagby, and M. G. Moharam.
Guided-mode resonances in planar dielectric-layer diffration grating.
Journal of the Optical Society of America A, 7:1470 – 1474, 1990.

[58] R. W. Wood.
Remarkable spectrum from a diffraction grating.
Philosophical Magazine, 4:396, 1902.

[59] R. W. Wood.
Anomalous diffraction gratings.
Physical Review, 48:928, 1935.

[60] A. Hessel and A. A. Oliner.
A new theory of Wood's anomalies on optical gratings.
Applied Optics, 4:1275, 1965.

[61] U. Fano.
Effects of configuration interaction on intensities and phase shifts.
Physical Review, 124:1866, 1961.

[62] M. Galli, D. Bajoni, M. Belotti, F. Paleari, M. Patrini, G. Guizzetti, D. Greace, M. Agio, L. C. Andreani, D. Peyrade, and Y. Chen.
Measurement of photonic mode dispersion and linewidths in silicon-on-insulator photonic crystal slabs.
IEEE Journal on Selected Areas in Comunications, 23:1402, 2005.

[63] A. Mekis, J. C. Chen, I. Kurland, S. Fan, P. R. Villeneuve, and J. D. Joannopoulos.
High transmission through sharp bends in photonic crystal waveguides.
Physical Review Letters, 77:3787–3790, 1996.

[64] J. D. Joannopoulos, P. R. Villeneuve, and S. Fan.
Photonic crystals: putting a new twist on light.
Nature, 386:143 – 149, 1997.

[65] S. Fan, P. R. Villeneuve, J. D. Joannopoulos, and E. F. Schubert.
High extraction efficiency of spontaneous emission from slabs of photonic crystals.
Physical Review Letters, 78:3294–3297, 1996.

[66] H. Yokoyama.
Physics and device applications of optical microcavities.
Science, 256:66–70, 1992.

[67] H. G. Park, S. H. Kim, S. H. Kwon, Y. G. Ju, J. K. Yang, J. H. Baek, S. B. Kim, and Y. H. Lee.
Electrically driven single-cell photonic crystal laser.
Science, 305:1444–1447, 2004.

[68] D. Kleppner.
Inhibited spontaneous emission.
Physical Review Letters, 47:233–236, 1981.

[69] S. Noda, M. Fujita, and T. Asano.
Spontaneous-emission control by photonic crystals and nanocavities.

Nature Photonics, 1:449–458, 2007.

[70] Y. Xu, R. K. Lee, and A. Yariv.
Quantum analysis and the classical analysis of spontaneous emission in a microcavity.
Physical Review A, 61:033807, 2000.

[71] A. F. Koenderink, M. Kafesaki, C. M. Soukoulis, and V. Sandoghdar.
Spontaneous emission in the near field of two-dimensional photonic crystals.
Optics Letters, 30:3210–3212, 2005.

[72] A. F. Koenderink, M. Kafesaki, C. M. Soukoulis, and V. Sandoghdar.
Spontaneous emission rates of dipoles in photonic crystal membranes.
Journal of the Optical Society of America B, 23:1196–1206, 2006.

[73] M. Gerken and Y. Nazirizadeh.
Spontaneous emission control for communications devices - a review.
Journal of Nanophotonics, 2:021795, 2008.

[74] D. H. Kim, C. O. Cho, Y. G. Roh, H. Jeon, Y. S. Park, J. Cho, J. S. Im, C. Sone, Y. Park, W. J. Choi, and Q. H. Park.
Enhanced light extraction from gan-based light-emitting diodes with holographically generated two-dimensional photonic crystal patterns.
Applied Physics Letters, 87:203508, 2005.

[75] U. Geyer, J. Hauss, B. Riedel, S. Gleiss, U. Lemmer, and M. Gerken.
Large-scale patterning of indium tin oxide electrodes for guided mode extraction from organic light-emitting diodes.
Journal of Applied Physics, 104:093111, 2008.

[76] C. Wiesmann, K. Bergenek, N. Linder, and U.T. Schwarz.
Photonic crystal LEDs - designing light extraction.
Laser & Photonics Review, 3:262 – 286, 2008.

[77] J. J. Wierer, A. David, and M. M. Megens.
III-nitride photonic-crystal light-emitting diodes with high extraction efficiency.
Nature Photonics, 3:163, 2009.

[78] E. Chow, A. Grot, L. W. Mirkarimi, M. Sigalas, and G. Girolami.
Ultracompact biochemical sensor built with two-dimensional photonic crystal microcavity.
Optics Letters, 29:1093 – 1095, 2004.

[79] M. R. Lee and P. M. Fauchet.

Two-dimensional silicon photonic crystal based biosensing platform for protein detection.
Optics Express, 15:4530 – 4535, 2007.

[80] J. L. Guo.
Nanoimprint lithography: Methods and material requirements.
Advanced Materials, 19:495 – 513, 2007.

[81] I. D. Block, L. L. Chan, and B. T. Cunningham.
Large-area submicron replica molding of porous low-k dielectric films and application to photonic crystal biosensor fabrication.
Microelectronic Engineering, 84:603 – 608, 2007.

[82] B. T. Cunningham, P. Li, B. Lin, and J. Pepper.
Colorimetric resonant reflection as a direct biochemical assay technique.
Sensors and Actuators B, 81:316 – 328, 2002.

[83] W. Zhang, N. Ganesh, I. D. Block, and B. T. Cunningham.
High sensitivity photonic crystal biosensor incorporating nanorod structures for enhanced surface area.
Sensors and Actuators B, 131:279 – 284, 2008.

[84] J. Li, Y. Wu, J. Fu, Y. Cong, J. Peng, and Y. Han.
Reversibly strain-tunable elastomeric photonic crystals.
Chemical Physics Letters, 390:285 – 289, 2004.

[85] Y. Ying, J. Xia, and S. H. Foulger.
Pressure tuning the optical transmission properties of photonic band gap composites.
Applied Physics Letters, 90:071110, 2007.

[86] T. F. Krauss, R. M. De La Rue, and S. Brand.
Two-dimensional photonic-bandgap structures operating at near-infrared wavelengths.
Nature, 383:699 – 702, 1996.

[87] Y. A. Vlasov, X.-Z. Bo, J. C. Sturm, and D. J. Norris.
On-chip natural assembly of silicon photonic bandgap crystals.
Nature, 414:289 – 293, 2001.

[88] S. Y. Chou, P. R. Krauss, and P.J. Renstrom.
Imprint lithography with 25-nanometer resolution.
Science, 272:85 – 87, 1996.

[89] K. Yee.

Numerical solution of initial boundary value problems involving maxwell's equations in isotropic media.
IEEE Transactions on Antennas and Propagation, 14:302 – 307, 1966.

[90] A. Taflove.
Application of the finite-difference time-domain method to sinusoidal steady state electromagnetic penetration problems.
IEEE Transactions on Electromagnetic Compatibility, 22:191 – 202, 1980.

[91] K. S. Kunz and R. J. Luebbers.
The Finite Difference Time Domain Method for Electromagnetics.
CRC Press, 1993.

[92] K. R. Umashankar and A. Taflove.
A novel method to analyze electromagnetic scattering of complex objects.
IEEE Transactions on Electromagnetic Compatibility, 24:397 – 405, 1982.

[93] J. Berenger.
A perfectly matched layer for the absorption of electromagnetic waves.
Journal of Computational Physics, 114:185 – 200, 1994.

[94] S. G. Johnson.
Notes on perfectly matched layers (pmls).

[95] `http://www.lumerical.com/`; visited on September 30th 2009.

[96] `http://ab-initio.mit.edu/wiki/index.php/Meep/`; visited on September 30th 2009.

[97] A. Taflove and S. C. Hagness.
Computational Electrodynamics: The finite-difference time-domain method.
Artech House, 2005.

[98] V. Pacradouni, W. J. Mandeville, A. R. Cowan, P. Paddon, Jeff F. Young, and S. R. Johnson.
Photonic band structure of dielectric membranes periodically textured in two dimensions.
Physical Review B, 62:4204, 2000.

[99] Yousef Nazirizadeh, Jürgen Müller, Ulf Geyer, Detlef Schelle, Ernst-Bernhard Kley, Andreas Tünnermann, Uli Lemmer, and Martina Gerken.
Optical characterization of photonic crystal slabs using orthogonally oriented polarization filters.
Optics Express, 16:7153–7160, 2008.

[100] M. Augustin, H.-J. Fuchs, D. Schelle, E.-B-Kley, S. Nolte, A. Tünnermann, R. Iliew, C. Etrich, U. Peschel, and F. Lederer.
Highly efficient waveguide bends in photonic crystal with a low in-plane index contrast.
Optics Express, 11:3284 – 3289, 2003.

[101] M. Augustin, H.-J. Fuchs, D. Schelle, E.-B-Kley, S. Nolte, A. Tünnermann, R. Iliew, C. Etrich, U. Peschel, and F. Lederer.
High transmission and single-mode operation in low-index-contrast photonic crystal waveguide devices.
Applied Physics Letters, 84:663 – 665, 2004.

[102] Y. Nazirizadeh, J. G. Müller, U. Geyer, U. Lemmer, and M. Gerken.
Direct observation of photonic modes in photonic crystal slabs.
Proc. ICTON, 2:72 – 75, 2008.

[103] K. B. Crozier, V. Lousse, O. Kilic, S. Kim, S. Fan, and O. Solgaard.
Air-bridged photonic crystal slabs at visible and near-infrared wavelengths.
Physical Review B, 73:115126, 2006.

[104] G. A. Turnbull, P. Andrew, M. J. Jory, W. L. Barnes, and I. D. W. Samuel.
Relationship between photonic band structure and emission characteristics of a polymer distributed feedback laser.
Physical Review B, 64:125122, 2001.

[105] G. A. Turnbull, P. Andrew, W. L. Barnes, and I. D. W. Samuel.
Photonic mode dispersion of a two-dimensional distributed feedback polymer laser.
Physical Review B, 67:165107, 2003.

[106] K. Sakai, E. Miyai, T. Sakaguchi, D. Ohnishi, T. Okano, and S. Noda.
Lasing band-edge identification for a surface-emitting photonic crystal laser.
IEEE Journal on Selected Areas in Comunications, 23:1335 – 1340, 2005.

[107] R. Harbers, P. Strasser, D. Caimi, R. F. Mahrt, N. Moll, D. Erni, W. Baechtold, B. J. Offrein, and U. Scherf.
Enhanced feedback and experimental band mapping of organic photonic-crystal lasers.
Journal of Optics A: Pure and Applied Optics, 8:273 – 277, 2006.

[108] A. R. Cowan, P. Paddon, V. Pacradouni, and J. F. Young.
Resonant scattering and mode coupling in two-dimensional textured planar waveguides.
Journal of the Optical Society of America A, 18:1160 – 1170, 2001.

[109] T. Ochiai and K. Sakoda.
Dispersion relation and optical transmittance of a hexagonal photonic crystal slab.
Physical Review B, 63:125107, 2001.

[110] R. Ferrini, R. Houdré, H. Benisty, M. Qiu, and J. Moosburger.
Radiation losses in planar photonic crystals: two-dimensional representation of hole depth and shape by an imaginary dielectric constant.
Journal of the Optical Society of America B, 20:469 – 478, 2003.

[111] G. von Freymann, W. Koch, D. C. Meisel, M. Wegener, M. Diem, A. Garcia-Martin, S. Pereira, K. Busch, J. Schilling, R. B. Wehrspohn, and U. Goesele.
Diffraction properties of two-dimensional photonic crystals.
Applied Physics Letters, 83:614 – 616, 2003.

[112] Y. Nazirizadeh, U. Geyer, Lemmer, and M. Gerken.
Spatially resolved optical characterization of photonic crystal slabs using direct evaluation of photonic modes.
IEEE/LEOS International Conference on Optical MEMs and Nanophotonics, pages 112 – 113, 2008.

[113] Y. A. Vlasov, M. Deutsch, and D. J. Norris.
Single-domain spectroscopy of self-assembled photonic crystals.
Applied Physics Letters, 76:1627 – 1629, 2000.

[114] M. Gerken, R. Boschert, R. Bornemann, U. Lemmer, D. Schelle, M. Augustin, E.-B. Kley, and A. Tünnermann.
Transmission measurements for the optical characterization of 2d-photonic crystals.
Proc. SPIE, 5963, 2005.

[115] S. Wang and S. Sheem.
Two-dimensional distributed-feedback lasers and their applications.
Applied Physics Letters, 22:460, 1973.

[116] M. Meier, A. Mekis, A. Dodabalapur, A. Timko, R. E. Slusher, J. D. Joannopoulos, and O. Nalamasu.
Laser action from two-dimensional distributed feedback in photonic crystals.
Applied Physics Letters, 74:7, 1999.

[117] H. Matsubara, S. Yoshimoto, H. Saito J. L. Yue, Y. Tanaka, and S. Noda.
GaN photonic-crystal surface-emitting laser at blue-violet wavelengths.
Science, 319:445 – 447, 2008.

[118] L. Chen, Z. X. Qiang, H. J. Yang, H. Q. Pang, Z. Q. Ma, and W. D. Zhou.

Polarization and angular dependent transmissions on transferred nanomembrane fano filters.
Optics Express, 17:8396 – 8406, 2009.

[119] B. H. Cheong, O. N. Prudnikov, E. Cho, H. S. Kim, J. Yu, Y. S. Cho, H. Y. Choi, and S. T. Shin.
High angular tolerant color filter using subwavelength grating.
Applied Physics Letters, 94:213104, 2009.

[120] N. Ganesh, W. Zhang, P. C. Mathias, E. Chow, J. A. N. T. Soares, V. Malyarchuk, A. D. Smith, and B. T. Cunningham.
Enhanced fluorescence emission from quantum dots on a photonic crystal surface.
Nature technology, 2:515, 2007.

[121] V. N. Astratov, M. S. Skolnick, S. Brand, T. F. Krauss, O. Z. Karimov, R. M. Stevenson, D. M. Whittaker, I. Culshaw, and R. M. De la Rue.
Experimental technique to determine the band structure of two-dimensional photonic lattices.
IEEE Proceedings-Optoelectronics, 145:398, 1998.

[122] V. N. Astratov, R. M. Stevenson, I. S. Culshaw, D. M. Whittaker, M. S. Skolnick, T. F. Krauss, and R. M. De La Rue.
Heavy photon dispersions in photonic crystal waveguides.
Applied Physics Letters, 77:178 – 180, 2000.

[123] M. Patrini, M. Galli, F. Marabelli, M. Agio, L. C. Andreani, D. Peyrade, and Y. Chen.
Photonic bands in patterned silicon-on-insulator waveguides.
IEEE Journal of Quantum Electronics, 38:885, 2002.

[124] K. Forberich, M. Diem, J. Crewett, U. Lemmer, A. Gombert, and K. Busch.
Lasing action in two-dimensional organic photonic crystal lasers with hexagonal symmetry.
Applied Physics B, 82:539 – 541, 2006.

[125] M. D. Whittaker and I. S. Culshaw.
Scattering-matrix treatment of patterned multilayer photonic structures.
Physical Review B, 60:2610, 1999.

[126] I. D. Block, N. Ganesh, M. Lu, and B. T. Cunningham.
A sensitivity model for predicting photonic crystal biosensor performance.
IEEE Sensors Journal, 8:274, 2008.

[127] L. Shi, P. Pottier, Y. A. Peter, and M. Skorobogatiy.

Guided-mode resonance photonic crystal slab sensor based on bead monolayer geometry.
Optics Express, 16:17962, 2008.

[128] R. Blankenburg, P. Meller, H. Ringsdorf, and C. Salesse.
Interaction between biotin lipids and streptavidin in monolayers: formation of oriented two-dimensional protein domains induced by surface recognition.
Biochemistry, 28:8214 – 8221, 1989.

[129] T. Sano and C. R. Cantor.
Intersubunit contacts made by tryptophan 120 with biotin are essential for both strong biotin binding and biotin-induced tighter subunit association of streptavidin.
Proceedings of the National Academy of Sciences, 92:3180, 1995.

[130] M. González and C. E. Argaraña amd G. D. Fidelio.
Extremely high thermal stability of streptavidin and avidin upon biotin binding.
Biomolecular Engineering, 16:67 – 72, 1999.

[131] T. Coussaert, A. R. Völkel, J. Noolandi, and A. P. Gast.
Streptavidin tetramerization and 2d crystallization: a mean-field approach.
Biophysical Journal, 80:2004 – 2010, 2001.

[132] M. R. Horton, C. Reich, A. P. Gast, J. O. Rädler, and B. Nickel.
Structure and dynamics of crystalline protein layers bound to supported lipid bilayers.
Langumuir, 23:6263 – 6269, 2007.

[133] S. Sekula, J. Fuchs, S. Weg-Remers, P. Nagel, S. Schuppler, J. Fragala, N. Theilacker, M. Franzreb, C. Wingren, P. Ellmark, C. A. K. Borrebaeck, C. A. Mirkin, H. Fuchs, and S. Lenhert.
Multiplexed lipid dip-pen nanolithography on subcellular scales for the templating of functional proteins and cell culture.
Small, 4:1785, 2008.

[134] H. M. Haake, A. Shuetz, and G. Gauglitz.
Label-free detection of biomolecular interaction by optical sensors.
Fresenius' Journal of Analytical Chemistry, 366:576 – 585, 2000.

[135] R. M. Ostroff, D. Hopkins, A. B. Haeberli, W. Baouchi, and B. Polisky.
Thin film biosensor for rapid visual detection of nucleic acid targets.
Clinical Chemistry, 45:1659 – 1664, 1999.

[136] W. D. Wright.

A re-determination of the trichromatic coefficients of the spectral colours.
Transactions of the Optical Society, 30:141 – 164, 1929.

[137] W. D. Wright and F. H. G. Pitt.
Hue-discrimination in normal colour-vision.
Proceedings of the Physical Society, 46:459 – 473, 1934.

[138] D. L. MacAdam.
Visual sensitivities to color differences in daylight.
Journal of the Optical Society of America, 32:247 – 274, 1942.

[139] S. K. Bhatia, L. C. Shriver-Lake, K. J. Prior, J. H. Georger, J. M. Calvert, R. Bredehorst, and F. S. Ligler.
Use of thiol-terminal silanes and heterobifunctional crosslinkers for immobilization of antibodies on silica surfaces.
Analytical Biochemistry, 178:408 – 413, 1989.

[140] I. D. Block, P. C. Mathias, S. I. Jones, L. O. Vodkin, and B. T. Cunningham.
Optimizing the spatial resolution of photonic crystal label-free imaging.
Applied Optics, 48:6567 – 6574, 2009.

[141] M. Bally, M. Halter, J. Vörös, and H. M. Grandin.
Optical microarray biosensing techniques.
Surface and Interface Analysis, 38:1442 – 1458, 2006.

[142] P. Domnanich, U. Sauer, J. Pultar, and C. Preininger.
Protein microarray for the analysis of human melanoma biomarkers.
Sensors and Actuators B, 139:2 – 8, 2009.

List of Figures

List of Tables

Index

Nomenclature

Symbols

$\vec{E}$	Electric field vector
$\vec{D}$	Dielectric displacement
$\vec{H}$	Magnetic field vector
$\vec{B}$	Magnetic flux
$\vec{j}$	Current
ρ	Charge
$\epsilon(\vec{r})$	Dielectric constant as a function of space
$\epsilon(\omega)$	Dielectric constant as a function of frequency
ϵ_∞	Dielectric constant at infinite frequency
μ	Permeability
$\vec{k}$	Wave vector
ω	Frequency
λ	Wavelength
c_0	Speed of light in vacuum
n_{core}	Core refractive index
$n_{cladding}$	Cladding refractive index
n_{eff}	Effective refractive index
d	Slab thickness
Λ	Periodicity
Q	Quality factor
Γ	Crystal direction
X	Crystal direction
M	Crystal direction
K	Crystal direction
θ	Illumination angle
ϕ	Azimuthal angle
R	Transducer response
R_0	Initial response
R_{max}	Maximum transducer response
k_a	Association rate constant
k_d	Dissociation rate constant
C_0	Concentration
τ_p	Photon lifetime
L_p	Photon propagation length

Abbreviations

1D, 2D, 3D	one-dimensional, two-nimesional, three-dimensional
TE	Transverse electric
TM	Transverse magnetic
rpm	Revolutions per minute
RIU	Refractive index unit
CCD	Charge-coupled Device
LED	Light emitting diode
DFB	Distributed feedback
VCSEL	Vertical-cavity surface-emitting lasers
FIB	Focused Ion Beam
SEM	Scanning electron microscope
AFM	Atomic force microscopy
ATR	Attenuated total reflection
PCS	Photonic crystal slab
GMR	Guide-mode resonances
QGM	Quasi guided modes
Q-Factor	Quality factor
SPP	Surface plasmon polariton
SPR	Surface plasmon resonance
SNR	Signal to noice ratio
FWHM	Full width at half maximum
PWE	Plane wave expansion method
GME	Guided-mode expansion method
FDTD	Finite-difference time-domain
PML	Perfectly matched layer
ABC	Absorbing boundary condition
PBC	Periodic boundary condition
PPBC	Bloch-periodic boundary condition
FFT	Fast Fourier transformation
CAD	Computer aided design
GUI	Graphical user interface
POCT	Point-of-care testing
ELISA	Enzyme-linked immunosorbent assay
CIE	International Commission on Illumination
IUPAC	International Union of Pure and Applied Chemistry

Materials

ITO	Indium tin oxide
Nb_2O_5	Niobium pentoxide
Ta_2O_5	Tantalum pentoxide
PMMA	Poly methyl methacrylate
PBS	Phosphate buffered saline
N-Biotinyl Cap-PE	1,2-dioleoyl-sn-glycero-3-phosphoethanolamine-N-(cap biotinyl)
DOPC	Phospholipid 1,2-dioleoyl-sn-glycero-3-phosphocholine

Acknowledgment

This thesis would not have been possible without the help of some people to whom this acknowledgment is addressed.

First of all I offer my sincerest gratitude to my advisors Prof. Dr. Martina Gerken and Prof. Dr. Uli Lemmer. Martina, thank you for giving me the opportunity to be part of your group. Your constant support throughout this work, whilst allowing me the room to work in my own way, made the realization of this work possible. Uli, thank you for giving me the opportunity to work at the Light technology Institute (LTI) as well as for many discussions, which were always an enrichment for my work.

I would like also to thank Prof. Dr. Juerg Leuthold for being co-examiner of my thesis.

I am grateful to Dr. Ulf Geyer for providing me with photonic crystal slabs at LTI, and Detlef Schelle and Dr. Ernst-Bernhard Kley for photonic crystal fabrication at the Institute of Applied Physics at the Friedrich-Schiller University in Jena.

I would like to thank Uwe Bog and Dr. Sylwia Sekula for their support and assistance in the biochemical part of this work.

For numerous useful comments and suggestions during preparation of this paper I especially thank Dr. Jost Adam, Uwe Bog, Dr. Ulf Geyer and Julian Hauß.

In my daily work I have been blessed with a friendly and cheerful group of colleagues at the LTI. Julian Hauß, Klaus Huska, Felix Glöckler and Boris Riedel thank you for an enjoyable working atmosphere.

I would like to thank also all my colleagues at the group Integrated Systems and Photonics (ISP) at the Christian-Albrechts-Universität zu Kiel. I really enjoined our scientific and non-scientific activities.

Finally, I would like to thank my parents and my family for their continuous moral support. Special thank goes to Mahsa, who had always infinite patience with me when I bored her with scientific discussions, which however were many times a source of new ideas to me.